THE COSMIC MYSTERY TOUR

NICHOLAS MEE

the COSMIC MYSTERY TOUR

A HIGH-SPEED JOURNEY THROUGH SPACE & TIME

OXFORD
UNIVERSITY PRESS

OXFORD
UNIVERSITY PRESS

Great Clarendon Street, Oxford, OX2 6DP,
United Kingdom

Oxford University Press is a department of the University of Oxford.
It furthers the University's objective of excellence in research, scholarship,
and education by publishing worldwide. Oxford is a registered trade mark of
Oxford University Press in the UK and in certain other countries

First Edition published in 2019

Impression: 1

Published in the United States of America by Oxford University Press
198 Madison Avenue, New York, NY 10016, United States of America

British Library Cataloguing in Publication Data
Data available

Library of Congress Control Number: 2018944463

ISBN 978–0–19–883186–0

DOI: 10.1093/oso/9780198831860.001.0001

Printed in Great Britain by
Bell & Bain Ltd., Glasgow

Preface

The Cosmic Mystery Tour will take you on a lightning trip around some of the greatest mysteries of our universe. Our trip will be interwoven with brief tales of the colourful characters who created modern science and lead us to the cutting-edge research of today.

In Part I we explore the laws that govern the cosmos. Physics is a spiritual quest to find deep meaning in the universe. Its goal is to provide a concise, but accurate description of the world that accounts for all the amazing features it contains.

In Part II we take a look at the history of the cosmos, study its geography and explore some of its architectural highlights, such as red giants, white dwarfs, neutron stars and the ultimate cosmic mysteries—supermassive black holes.

In Part III we consider the possibility that life might exist elsewhere. We will explore the cosmos from the outer fringes of science fiction to the ongoing search for alien intelligence.

I am happy to be your tour guide.

Acknowledgements

I would like to thank all the staff at the Science Museum in London for being so welcoming and for their help in making the books-signing events such a great success.

Thank you to everyone who has joined my mailing list at http://www.quantumwavepublishing.co.uk, and thank you for all the feedback on my blog articles on the Quantum Wave website.

Thank you to John Eastwood for all his help at Astrofest, New Scientist Live and numerous other book-signing events.

Many thanks to Quentin Williams, Mark Sheeky, Laura Dewar, Donna Barlow, Carl Larsen and Julius Mazzarella for reading through early drafts of various chapters and for their helpful feedback. Special thanks to Heidi Chapman, Mhairi Gray and Debra Nightingale for reading through the whole manuscript and for their valuable comments and encouragement. And a further thanks to Mhairi Gray for compiling the index.

Thank you to Nick Manton for inviting me to co-author *The Physical World*, which helped to pave the way for *The Cosmic Mystery Tour*. Thank you to Sonke Adlung, Ania Wronski and everyone at Oxford University Press for turning my manuscript into such a beautiful book.

Many thanks to Angie for putting up with another book project, and thank you to my parents for all their encouragement and support.

And a special thanks to Phil Withers for telling the funniest stories I have ever heard.

Contents

PART I

THE LAWS OF THE COSMOS

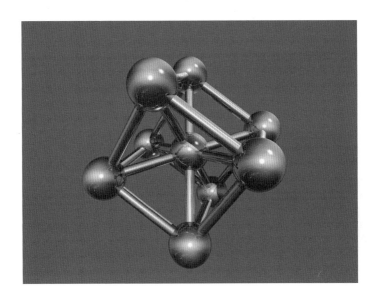

The Path to Immortality

It is 1697. We enter a dimly lit tavern in one of the less inviting districts of London. Huddled in the shadows is a man in a loose cloak sitting expectantly with his accomplice at a small table. He has long grey hair, a sharp nose and a determined look in his piercing eyes. This is the Warden of the Royal Mint and he is hoping to interrogate some of the coiners and counterfeiters who, along with many other assorted cutthroats and villains, haunt this alehouse and its environs.

Isaac Newton has been appointed warden—a position worth £300 annually—just the previous year. The modernization of the currency is underway, hand-stamped coins remain in circulation, but the mint is now mechanized with coins of silver and gold being smelted and stamped in the Tower of London on an industrial scale (Figure 1). Each year the Royal Mint pays £700 just to remove the dung generated by its horse-powered machinery. Newton has declined the accommodation adjacent to the smoke, stench and noise of the factory close by the army barracks within the walls of the royal fortress, and has made his own arrangements in the heart of swinging London. Even so he engages in little of the high life the capital has to offer. Newton is no enthusiast for the arts, alluding to classical sculptures as 'stone dolls' and describing poetry as 'a kind of ingenious nonsense'. He does visit the opera *once* in what is the golden age of English opera, the era of Henry Purcell. He later recalls: 'I heard the first act with pleasure, the second act stretched my patience and in the third act I ran away.'

The Cosmic Mystery Tour. Nicholas Mee, Oxford University Press (2019). © Nicholas Mee.
DOI: 10.1093/oso/9780198831860.001.0001

Figure 1 The Tower of London.

Newton's interest in the mint may have emerged from his alchemical investigations. He probably did not realize his duties would extend to the prosecution of coiners, clippers and counterfeiters. Soon after taking on the role, he wrote to the Treasury requesting that a duty 'so vexatious & dangerous' not be required of him any longer. Their response was simply to insist that Newton meet his obligations. With little option but to comply, Newton took to the task with the single-minded determination he applied to all his endeavours.

A Visit to the Underworld

We find it hard to imagine the father of modern science descending into the underworld vice dens, alehouses and taverns, and the notorious Newgate prison gathering evidence for the conviction and execution of London's coiners. It invokes an image common to the world's mythologies of the hero descending into an underworld realm of horror and depravity where, after overcoming the most arduous ordeals and ultimately death itself, the hero returns with heightened knowledge and self-awareness.

The heroes Gilgamesh, Theseus, Heracles, Odysseus and Aeneas all undertook such a journey and their mythical success was proof of their extraordinary powers and the key to their immortality. Newton's immortality had been established in the much quieter and more genteel surroundings of Trinity College, Cambridge, where a decade earlier in 1687 he published the book that more than any other would forge the modern scientific world. This book, *Philosophiæ Naturalis Principia Mathematica* (Mathematical Principles of Natural Philosophy) is usually known as *The Principia* (pronounced with a hard 'c').

Isaac Newton entered Trinity College (Figure 2) in 1661 as an eighteen-year-old student. In August 1665 the university closed due to an outbreak of plague and Newton returned home to Woolsthorpe Manor in Lincolnshire. Newton remained there for the next two years and it was during this period he developed his early ideas that would lead to a complete transformation of science and the wider world. There is a famous and memorable image of the young Newton musing on the force of gravity after seeing an apple fall in the garden of the manor house. Although this sounds an unlikely story, it was told by Newton himself. Newton lived his last seven years with his niece Catherine Barton and her husband, John Conduitt, who also assisted Newton at

Figure 2 Great Court, Trinity College, Cambridge.

the Royal Mint. Conduitt recorded the following account of Newton's story:

> In the year 1666 he retired again from Cambridge to his mother in Lincolnshire. Whilst he was pensively meandering in a garden it came into his thought that the power of gravity (which brought an apple from a tree to the ground) was not limited to a certain distance from earth, but that this power must extend much further than was usually thought. Why not as high as the Moon said he to himself & if so, that must influence her motion & perhaps retain her in her orbit, whereupon he fell a calculating what would be the effect of that supposition.

Newton realized that if we throw an apple it will follow an arc as it is drawn towards the ground by gravity. We can imagine firing an apple or a cannonball from a mountain top. The faster we propel the cannonball the further it will travel before hitting the ground. Newton reasoned that if propelled with sufficient velocity the cannonball would circle the Earth completely, like the Moon,

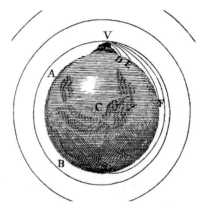

Figure 3 Newton's illustration from page 6 of his *Philosophiæ Naturalis Principia Mathematica*, Volume 3, *De mundi systemate* (On the system of the world).

without ever touching the ground—it would be in orbit. Newton illustrated this idea in *The Principia* with the picture shown in Figure 3. We are so familiar with the notion that apples are pulled to the ground by the same force that holds the planets in orbit, it is hard to conceive of a time when this was not common knowledge. But prior to the Newtonian revolution, less than 350 years ago, the connection between the fall of an apple and the celestial dance of the planets was completely unknown.

The Scientific Revolution

Newton offered far more than an analogy and a convincing argument. He produced a complete system of mechanics and gravity that could be used to calculate how the material world operates. Newton's ideas were applied far and wide by subsequent generations and became the foundation for the entire scientific enterprise of understanding the universe. Newton's theories reigned supreme for over 200 years (Figure 4).

It was not until the early years of the twentieth century that modifications to Newton's theories were found to be necessary. We now know that quantum theory is required when analysing the very small, special relativity when objects

Figure 4 Saturn's beautiful rings held within the planet's gravitational embrace.

are moving close to the speed of light, and general relativity when considering the gravitational effects of very massive bodies.

Warping Space and Time

In 1915, Albert Einstein devised a theory of gravity—general relativity—that works even better than Newton's. Einstein did away with Newton's force of attraction between massive bodies. He proposed instead that each massive body warps space and time in its vicinity and this affects the path of any object moving close by, including light. Although these descriptions sound completely different, the predictions of the two theories are very similar unless the bodies are extremely massive.

Rather remarkable conclusions follow from Einstein's theory when applied to intense gravitational environments. General relativity implies that massive objects bend the path of light and objects such as giant elliptical galaxies warp space so much they act like enormous gravitational lenses. Figure 5 shows an example where the light from a distant galaxy is warped into a circle by the intervening giant elliptical galaxy seen within the ring. Another consequence of general relativity is the existence of black holes—objects whose gravitational attraction is so severe that even light cannot escape. General relativity also predicts that incredibly violent events, such as black hole collisions and mergers, generate ripples in the fabric of space known as gravitational waves. All these outlandish predictions are now known to be true.

Newtonian gravity works incredibly well. It is perfectly adequate in almost all circumstances. In the vicinity of the Earth, the differences between Newtonian gravity and general relativity are tiny, so we might expect them to be irrelevant in every day life.

Figure 5 Gravitational lens LRG 3-757 discovered in data from the Sloan Digital Sky Survey. The diffuse sphere within the arc is a giant elliptical galaxy. The arc itself is the distorted image of a more distant galaxy whose light has bent around the intervening giant elliptical.

It is quite surprising, therefore, that we now daily use a technology that relies on general relativity for its accuracy. Gravitational time distortion is built into the Global Positioning System (GPS) used for satellite navigation. GPS could not function for more than a few minutes if the effects of general relativity were not incorporated into the system.

Sir Isaac Newton

Newton succeeded in his crackdown on abuse of the British coinage and was personally responsible for 28 convictions. He was promoted to Master of the Royal Mint in 1699, and six years later when knighted by Queen Anne it was for his work at the Royal Mint rather than his scientific achievements. In 2017, the Royal Mint issued a 50p coin to mark the 375th anniversary of Newton's birth (Figure 6).

Figure 6 Commemorative Newton 50p coin.

Newton has only one rival to the title of greatest physicist of all time. Yet, there is much that Newton could never have known. By the end of this book your knowledge will surpass Newton's in many ways. You will even know where and how gold is created.

The Rosetta Stone and Quantum Waves

Thomas Young has been described as *the last man who knew everything* (Figure 7). He was certainly a remarkable scientist with wide-ranging interests. Born in 1773 in the small Somerset town of Milverton, by the age of 14 he was already fluent in Greek and Latin and over ten other languages both ancient and modern.

Figure 7 The great scientist and linguist Thomas Young.

The Cosmic Mystery Tour. Nicholas Mee, Oxford University Press (2019). © Nicholas Mee.
DOI: 10.1093/oso/9780198831860.001.0001

Cracking the Code

Much later in life, Young became fascinated by one of the great puzzles of the age—the decipherment of Egyptian hieroglyphics (Figure 8). The Rosetta Stone had been discovered in 1799 by a soldier in Napoleon's army, and following the French defeat in the Battle of the Nile, the famous engraved stone was shipped back to England. It was immediately recognized as the key to unlocking the ancient mystery of Egyptian writing, as the hieroglyphic inscription is repeated in a well-understood language—Ancient Greek. Young made the first significant progress in cracking the hieroglyphic code in 1814 and paved the way for the full decipherment by the French linguist Jean-Francois Champollion in the 1820s.

Figure 8 Thutmosis III cartouches on Hatshepsut's Temple at Deir el-Bahri.

Waves in a Sea of Light

But Young's most important achievement was to establish the wave-like nature of light. We take for granted today that light behaves like a wave, but the great Isaac Newton, whose influence

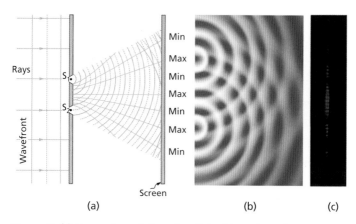

Figure 9 (a) Young showed that when light shines through two narrow slits in a screen, the overlapping light waves spreading out from the slits interfere. In some regions, the waves add up; in other regions they cancel out. (b) The interference pattern formed by the overlapping light waves looks very similar to the ripples on a pond. (c) When the light is projected onto a screen we see a cross section of the full interference pattern, a series of bright and dark regions known as interference fringes.
Credit: This image has been provided courtesy of PASCO Scientific.

dominated the physics of Young's day, believed that light was composed of particles. Young showed in a number of experiments that light forms interference patterns that are very similar to those produced by water waves on a pond and thereby demonstrated quite conclusively that light has wave-like properties (Figure 9).

The Language of Quantum Waves

This was certainly not the end of the story. The year 1905 saw one of the most surprising developments in the history of science when Einstein showed that Young and Newton were both correct. Light is composed of fundamental particles known as photons. So, somehow light simultaneously has the characteristics of both waves and particles. This amazing discovery heralded a total transformation in our understanding of the universe. And there was more, much more.

Louis de Broglie (1892–1987), whose surname is pronounced 'de Broy', was a French aristocrat who would eventually inherit the title of 7th duc de Broglie. Originally a history student, he later turned his attention to maths and physics. De Broglie presented an astonishing idea in his PhD thesis of 1924. He proposed that electrons might also have a wave-like nature and the higher their energy, the shorter their wavelength, just like photons. He went on to suggest that this should apply to all particles, meaning that not only light, but also matter is composed of constituents that

Figure 10 Quantum Man by Julian Voss-Andreae, a sculptor whose graduate research was in quantum physics. He participated in ground-breaking experiments showing that even large molecules such as buck-minsterfullerene produce interference patterns. His sculptures are designed to open our minds to the bizarre world of quantum mechanics.

somehow have the characteristics of both particles and waves. Such strange behaviour had already been observed in light, but never in matter. Incredibly, this monstrous proposal turned out to be correct. It was the final straw that brought the classical world crashing down and ushered in the development of modern quantum mechanics. The following decade saw an explosion in the application of quantum theory throughout physics and chemistry. The wave nature of electrons was demonstrated experimentally in 1927 and two years later De Broglie was awarded the Nobel Prize in Physics.

Wave–particle duality, as it is known, lies at the heart of the modern understanding of matter and the forces that hold it together. Deep down the concrete and familiar world that we all know and love is built on the complex shifting patterns formed by myriads of interfering quantum waves (Figure 10).

Seeing with Electron Waves

By the mid-1930s the discovery of electron waves had spawned a very important technological application—the electron microscope. These instruments create images using electron beams rather than visible light and achieve much greater magnification than optical microscopes as the wavelength of the electrons is much shorter than the wavelength of visible light. Optical microscopes magnify images by up to around two thousand times, whereas electron microscopes may attain magnifications of over one million times (Figure 11).

Following the devastation of the Second World War, De Broglie championed the idea of promoting peaceful international collaboration by establishing a trans-European laboratory where the newly emerging nuclear technologies could be developed and exploited. At the European Cultural Conference held in Lausanne in December 1949 De Broglie put forward the first official proposal

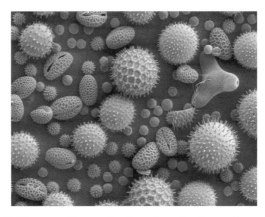

Figure 11 Electron micrograph of a variety of pollen grains.

for the creation of such a research centre. Following support at the fifth UNESCO General Conference, held in Florence in June 1950, the European Organization for Nuclear Research known as CERN was founded in 1954. It now has twenty-two member states. The CERN laboratory straddles the Franco-Swiss border near

Figure 12 The innards of the Compact Muon Solenoid (CMS) detector at the Large Hadron Collider.

Geneva. It is home to the world's largest particle accelerator—the Large Hadron Collider (Figure 12).

The Crazy World of Quantum Mechanics

The crazy world of quantum mechanics is a fundamental part of the world of physics. Many mysteries remain about its true meaning, but nonetheless it lies at the heart of a wide range of modern technologies. Lasers, superconducting magnets, holograms, computer processors and a whole host of other semiconductor devices all rely on quantum principles. But this is only the tip of a very large iceberg. The application of quantum theory will deliver ever more exotic technologies as we move into the future.

We're Having a Field Day!

In the 1630s René Descartes proposed an imaginative explanation for the motion of the planets. In a work known as *The World*, he argued that the existence of a void or vacuum is impossible and therefore space must be filled with some sort of fluid. Descartes suggested the planets are carried around the Sun by vortices in this fluid. Prior to Newton's theory of gravity this was the leading explanation for the planetary orbits, so when Newton (Figure 13) published his own theory in *The Principia* he went to great lengths to show that Descartes' theory could not work.

Newton's explanation of gravity was a great triumph. It gave an accurate account for the motion of the planets and much else besides. The clockwork of the heavens was explained by invoking long-range interactions between the Sun and planets, and indeed all massive bodies. The theory did not, however, include any apparent mechanism to transmit the force. Despite the extraordinary success of Newton's ideas, this action-at-a-distance was heavily criticized by philosophers such as Gottfried Leibniz.

Newton agreed. He admitted it was difficult to see how a force could operate between two bodies that are not in contact. In a letter to Richard Bentley in 1693 he wrote:

> It is inconceivable that inanimate matter should, without the mediation of something else, which is not material, operate upon, and affect other matter without mutual contact …
>
> That gravity should be innate, inherent and essential to matter, so that one body may act upon another at a distance through a

The Cosmic Mystery Tour. Nicholas Mee, Oxford University Press (2019). © Nicholas Mee.
DOI: 10.1093/oso/9780198831860.001.0001

Figure 13 Isaac Newton at the age of 46 in 1689, painted by Godfrey Kneller.

vacuum, without the mediation of anything else, by and through which their action and force may be conveyed from one to another, is to me so great an absurdity that I believe no man who has in philosophical matters a competent faculty of thinking can ever fall into it.

Centuries would pass before this enigma was resolved.

Faraday's Fields

Michael Faraday unlocked the secrets of electricity and magnetism in a long series of experiments in the first half of the nineteenth century. One of Faraday's great innovations was the idea of mapping out electric and magnetic forces throughout space. These maps plot a representative collection of lines showing the direction of the electric force experienced by a hypothetical test charge at each point in space, or the magnetic force experienced

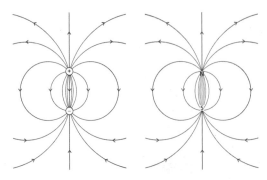

Figure 14 Left: electric field close to two charges. Right: magnetic field close to a bar magnet.

by a test magnetic pole. The strength of the force at each point is indicated by the density of the lines. Faraday referred to these maps as electric and magnetic *fields*.

A positive charge is repelled by another positive charge and attracted to a negative charge. The blue force lines on the left in Figure 14 indicate the direction of the resultant electric force that would be experienced by a positive test charge at each point in space due to the attraction to the negative charge and the repulsion from the positive charge. Similarly, the red lines on the right show the overall magnetic force experienced by a test north pole due to its repulsion from the north magnetic pole and its attraction to the south magnetic pole.

Although electric and magnetic fields are depicted separately, a changing electric field generates a magnetic field and a changing magnetic field generates an electric field, so it is natural to consider them as parts of a single electromagnetic field.

Faraday agonized over the true meaning of the fields and concluded they must have a real physical existence, but he was strongly criticized for this. When James Clerk Maxwell (Figure 15) formulated a mathematical theory of electromagnetism encapsulating

Figure 15 James Clerk Maxwell.

the results of Faraday's experiments their nature was greatly clarified.

Maxwell's equations show that waves are produced whenever an electromagnetic field is disturbed and these waves carry energy and momentum, so the field is not just an accounting tool; it really does have an independent existence. The waves are perpendicular oscillations of electric and magnetic forces propagating through space, as shown in Figure 16. The changing electric field generates the magnetic field and the changing magnetic field simultaneously generates the electric field, so the oscillating electromagnetic waves are self-perpetuating.

What is most remarkable is that we have always been aware of these waves; they are what we call light. Following experiments with polarized light, Faraday had suspected light to be an electromagnetic phenomenon. Tragically, when Maxwell visited Faraday to tell him he was correct, Faraday was close to death and already too ill to understand what Maxwell was saying.

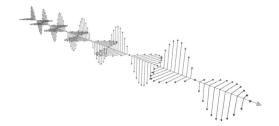

Figure 16 Electromagnetic wave. As shown in Figure 14, the blue arrows indicate the electric field and the red arrows indicate the magnetic field.

Visible light consists of rapid, high energy vibrations, with blue light vibrating more rapidly and with higher energy than red light. But Maxwell's calculations showed that it should be possible to generate electromagnetic vibrations of all frequencies (Figure 17).

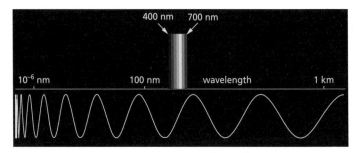

Figure 17 In the electromagnetic spectrum short wavelength vibrations have higher energy than long wavelength vibrations.

Radio Ripples

Maxwell's theory implies that when an oscillating current passes through a wire, electrons in the wire oscillate and as they accelerate they generate ripples in the electromagnetic field. In other words, they radiate electromagnetic waves.

Long wavelength vibrations are known as radio waves, a name derived from *radius*, Latin for the spoke of a wheel. The radio waves are emitted in all directions and move outwards forming a spherical wavefront. (This is analogous to the ripples produced when a stone is dropped into a pond.) When the wavefront impinges on a metal wire or *aerial*, the radio waves accelerate electrons in the aerial and this generates an oscillating current. The spherical wavefront will have spread out so the signal will be much weaker and may need amplification.

This is the basis for an experiment performed by Heinrich Hertz in 1887. Hertz generated radio waves at one end of his laboratory

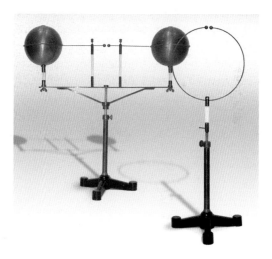

Figure 18 Some of Hertz's equipment.

and detected them at the other end. Some of his equipment is shown in Figure 18.

Quantum Waves

The discovery of quantum mechanics in the twentieth century shed further light on the meaning of fields. From a quantum perspective, electromagnetic waves are composed of particles known as photons. They are not particles in the classical billiard ball sense, more like discrete bits of wave. They are sometimes referred to as *quanta* or even *wavicles,* as everyday concepts such as particles or waves don't fully capture their properties.

Quantum field theory gives us a remarkable picture of how forces work at the level of particle interactions. For each type of fundamental particle a separate field is presumed to permeate space. Photons are fundamental vibrations of the electromagnetic field, electrons are vibrations of the electron field and so on. Crucially, these fields are not independent. For instance, the particles that are electrically charged, such as electrons, are those whose fields couple to the electromagnetic field. This produces a force between charged particles due to the exchange of pulses in the electromagnetic field. These pulses are photons.

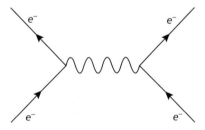

Figure 19 Time flows upwards in the diagram. The solid lines represent two electrons that approach each other, exchange a photon, as represented by the wavy line, then recede from each other.

Physicists represent such interactions pictorially in Feynman diagrams. The diagram in Figure 19 shows the exchange of a single photon between two electrons. This is the simplest, but most important, diagram representing the interaction between two electrons. Other diagrams would involve more particles.

It is very satisfying to arrive at this quantum description of particle interactions. Electrons mutually repel, as they are negatively charged, and this force causes two electrons to accelerate away from each other. Accelerating charged particles, such as electrons, generate electromagnetic waves and conversely electromagnetic waves accelerate any electrons they encounter. What could be more natural than to describe the change in energy and momentum of two interacting electrons as due to the exchange of pulses of electromagnetic waves between them?

Feynman diagrams provide a bridge between the mathematics of quantum field theory and our conceptual imagination of what is actually going on when particles interact. Niels Bohr, one of the architects of quantum theory, was critical when Richard Feynman (Figure 20) introduced them as they appear to hide some of the

Figure 20 Richard Feynman with his family in front of the van he decorated with examples of his diagrams.

mystery of what happens at the quantum level. Bohr later apologized when Feynman won the Nobel Prize in Physics in 1965.

Einstein Solves Newton's Puzzle

Einstein's greatest insight was to solve Newton's dilemma by banishing action-at-a-distance from gravity using the field concept. Einstein's general relativity is a classical theory of gravity, just as Maxwell's theory is a classical theory of electromagnetism. It predicts the existence of gravitational waves. These are continuous classical waves in the fabric of space. In the next chapter we will take a closer look at gravitational waves and the amazing instruments now being used to detect them.

We are still some way from a quantum theory of gravity. Such a theory would describe gravity as due to the exchange of quantum pulses of gravitational waves—hypothetical particles known as gravitons. These would be the gravitational equivalent of photons. However, gravity is so weak compared to electromagnetism that the prospects for ever detecting individual gravitons are remote. I have my doubts about whether it is even possible in principle.

Cosmic Ripples

Michael Faraday (Figure 21) transformed our understanding of the physical world when he realized that electromagnetic forces are carried by a field permeating the whole of space. This idea was formalized by James Clerk Maxwell who constructed a unified theory of electromagnetism in which beams of light are undulations in the electromagnetic field. Maxwell's theory implies that visible light is just one part of the electromagnetic spectrum and Hertz confirmed this experimentally by generating and detecting radio waves. The invention of radio followed, along with television, radar, mobile phones and many other applications. Electromagnetic waves are emitted whenever electrically charged objects, such as electrons, are shaken.

Figure 21 Michael Faraday commemorated on a £20 bank note.

The Cosmic Mystery Tour. Nicholas Mee, Oxford University Press (2019). © Nicholas Mee.
DOI: 10.1093/oso/9780198831860.001.0001

The Gravitational Field

When Einstein formulated his new theory of gravity—general relativity—he aimed to explain gravity as a theory of fields. In this he was successful. Remarkably, it turned out the appropriate field is spacetime itself.

In general relativity, spacetime is analogous to the electromagnetic field and mass is analogous to electric charge. One implication of the theory is that vigorously whirling large masses around will generate gravitational waves, and because gravity is described as the warping and curvature of spacetime, these gravitational waves are simply ripples in the fabric of space (Figure 22).

Detecting electromagnetic waves is easy. We do it whenever we open our eyes, turn on the television, use Wifi, or heat a cup of tea in a microwave oven. Detecting gravitational waves is rather more difficult, because gravity is incredibly weak compared to the electromagnetic force.

We live in an environment where gravity is very important and this gives a false impression of its strength. But it takes a planet-sized amount of matter pulling together for gravity to have a significant effect, and even then it is easy to pick up metal objects

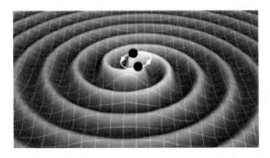

Figure 22 Schematic illustration of gravitational waves emitted by a binary black hole system.

with a small magnet, defying the gravitational attraction of the entire Earth.

Gravity is so weak that even shaking huge masses generates barely the tiniest gravitational ripple. Only the most violent cosmic events produce waves that could conceivably be detected; these include supernova explosions, neutron star collisions and black hole mergers. Any instrument sensitive enough to detect them must be able to measure changes in the distance between two points several kilometres apart by less than a thousandth of the size of a proton or about a billionth of the size of an atom. Incredibly, such instruments now exist.

Detecting the Ripples

In the centenary year of Einstein's general relativity, gravitational wave researchers achieved their first success. It had taken decades to develop the technology to build the Laser Interferometer

Figure 23 The LIGO facility in Livingston, Louisiana.

Gravitational-Wave Observatory (LIGO), consisting of two facilities 3000 kilometres apart in the United States, at Hanford, Washington, and Livingston, Louisiana (Figure 23). (Two well-separated detectors are required to distinguish true gravitational wave events from the inevitable local background disturbances.)

The facilities are L-shaped with two perpendicular 4-kilometre arms housed within an ultrahigh vacuum. A laser is directed at a beam-splitter, sending half the beam down each arm. The light travels 1600 kilometres, bouncing back and forth 400 times between two mirrors in each arm, before the two half-beams are recombined. The apparatus is designed so that the recombined half-beams completely cancel, with the peaks in the light waves of one beam meeting the troughs in the other, and no light passes to the light detector. Whenever a gravitational wave ripples through the apparatus, however, the lengths of the arms alter very slightly, so the distances travelled by the half-beams change and their phases shift (by much less than a single wavelength). There is no longer perfect cancellation and some light arrives at the light detector, as shown in Figure 24. The sensitivity of LIGO is

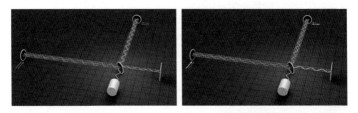

Figure 24 Schematic diagram of the LIGO gravitational wave detector. Left: The half-beams are exactly out of phase when they recombine and cancel perfectly, so no signal reaches the photodetector. Right: The end mirrors are displaced slightly, so the distances travelled by the half-beams changes. They are no longer exactly out of phase when recombined and now a signal does reach the photodetector.

extraordinary, as it must be if there is any chance of detecting gravitational waves.

Extreme Violence in the Depths of Space

The upgraded LIGO programme known as Advanced LIGO was scheduled to begin on 18 September 2015. Four days before the official start something wonderful happened. An unmistakable and identical signal was measured by the detectors in Hanford and Livingston within a few milliseconds of each other (Figure 25).

Researchers have studied computer models of black hole mergers and other violent cosmic processes so they can recognize the signatures of events detected by LIGO. According to the models, binary black holes produce a continuous stream of gravitational waves that drains energy from the binary system and the black holes gradually spiral together. In the final moments of

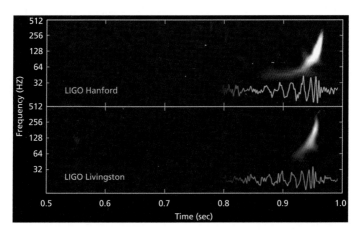

Figure 25 The first ever gravitational wave signal detected by the LIGO observatories.

inspiral the amplitudes of the gravitational waves increase dramatically. Initially, the newly merged black hole is rather asymmetrical, but it rapidly settles down with a final blast of gravitational waves known as the *ring-down*.

Much information has been extracted from this first brief signal detected by LIGO. It came from an event 1.3 billion light years away and was detonated by two merging black holes during their final inspiral and ring-down. The masses of the black holes are deduced to be 29 and 36 solar masses and they coalesced into a rapidly spinning black hole of 62 solar masses. What is truly staggering is that during the merger three times the mass of the Sun was converted into pure energy in the form of gravitational waves. This huge energy now forms a tiny gravitational ripple propagating outwards in an expanding sphere of 1.3 billion light years radius. The gravitational wave may be an incredibly small ripple, but the total energy it carries is enormous.

This was the first ever detection of a binary black hole system and the most direct observation of black holes ever made. It also confirmed that gravitational waves travel at the speed of light, as expected.

The 2017 Nobel Prize in Physics was awarded to three American physicists Kip Thorne, Barry Barish and Rainer Weiss, who played leading roles in the development of LIGO. The era of gravitational wave astronomy has now begun.

Lovely LISA

One of the amazing ideas to emerge from Einstein's theory of general relativity was the possibility of gravitational waves rippling their way across the cosmos. It took a century to verify this prediction. Their existence was finally confirmed by the Laser Interferometer Gravitational-Wave Observatory (LIGO) in September 2015, as described in the previous chapter.

What's in a Name?

LIGO is now detecting a new gravitational wave signal every few months. The first four signals were all due to black hole mergers

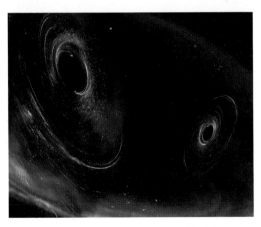

Figure 26 Artist's impression of a close binary black hole system based on GW170104.

The Cosmic Mystery Tour. Nicholas Mee, Oxford University Press (2019). © Nicholas Mee.
DOI: 10.1093/oso/9780198831860.001.0001

in the distant universe. Figure 26 is an artist's impression of the system that produced the third of these signals, detected on 4 January 2017. The black holes were 32 and 19 times the mass of the Sun and were spinning in different planes, as depicted in the illustration which shows them just before their merger. The signal has been named GW170104. Guess why?

The fifth signal detected by LIGO was rather different. We will investigate further in a later chapter, 'Cosmic Spacequakes', in Part II.

Triangulating Space

The detection of gravitational waves by LIGO is an incredible technological achievement. The European Space Agency (ESA) is planning to go one better by putting a gravitational wave detector in

Figure 27 Schematic representation of LISA.

space. It is known as the Laser Interferometer Space Antenna (LISA). There will be three spacecraft orbiting the Sun in a triangular formation, consisting of a mother and two daughter craft, each separated by a distance of 2.5 million kilometres (Figure 27). They will form a precision interferometer with lasers monitoring the distances between the mother and daughter craft. A passing gravitational wave will change these distances very slightly and this will be detected by the interferometer.

Paving the Way for LISA

The LISA Pathfinder mission launched in December 2015 as a stepping stone to the LISA mission was devised to test the technology and demonstrate the feasibility of constructing an interferometer in space. LISA Pathfinder released two cubic test masses to float freely within the spacecraft and used its laser interferometer to measure their separation, as shown in Figure 28. It then monitored their positions to an unprecedented accuracy of less than one hundredth of a nanometre.

The goal was to show that test masses can be shielded from any stray internal and external forces and maintained in a state of almost perfect free fall. Remarkably, the spacecraft avoids any contact with the test masses contained within its structure by

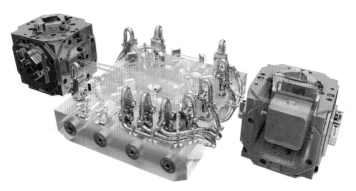

Figure 28 The LISA Pathfinder mission released two cubic test masses to float freely in solar orbit while their separation was monitored by a precision interferometer.

sensing their motion and adjusting its own position using micro-thrusters to compensate. This is essential if the sensitivity of the interferometer is not to be destroyed by the inevitable perturbations the spacecraft will suffer. These arise from a number of sources including stray gas molecules within the spacecraft, the solar wind and micro-meteoroid impacts.

In the Pathfinder mission the test masses are located just 40 centimetres apart, whereas the three LISA craft will be separated by millions of kilometres. The LISA interferometer will measure their separation just as accurately as the Pathfinder mission, so its sensitivity will scale up in proportion to its increased size. ESA announced in June 2017 that the technology trialled by LISA Pathfinder has performed beyond expectations, which means it will definitely be sensitive enough to detect gravitational waves when deployed by the LISA mission.

What Will We See?

LISA will greatly enhance our ability to study gravitational waves. It will detect signals invisible to LIGO and other ground-based gravitational wave detectors, as it will be sensitive to gravitational waves with much longer wavelengths produced by much larger systems. Although the black holes that merged during the GW170104 event were very massive, they were only 190 and 115 kilometres in diameter, with the merged black hole around 280 kilometres in diameter. These are very small objects by cosmic standards.

LISA will detect gravitational wave signals emanating from tightly bound binary systems containing two compact objects that may be white dwarfs, neutron stars or black holes orbiting each other prior to their merger. For instance, a binary black hole system such as GW170104 would be detected weeks or even months before the merger event. This will enable the position of the binary system to be located in the sky and the time of merger to be accurately predicted, which will greatly aid the visual identification of the merger event.

Supermassive Black Holes

There is a supermassive black hole of four million solar masses at the centre of our galaxy (Figure 29). (We will take a closer look at black holes in a later chapter: 'Doctor Atomic and the Black Hole'.) Indeed, most, if not all, galaxies are thought to harbour a monster such as this within their core. LISA will be able to detect these beasts devouring nearby stars. It will also detect mergers of supermassive black holes. Such extremely violent and spectacular events must occasionally happen somewhere in the universe. We can look forward to finding out much more about them.

The creation and early growth of supermassive black holes is still not well understood. LISA should detect their birth pangs and provide important clues to how they formed and their relationship to quasars in the early universe. LISA will also help to

Figure 29 An artist's impression of a supermassive black hole.

improve models of the immediate aftermath of the Big Bang and the very early universe. As an important bonus, LISA will add to our knowledge of fundamental physics by providing stringent new tests for general relativity.

LISA is scheduled for launch in 2034 as part of ESA's Cosmic Vision programme.

Animated Atom Boy

The idea of atoms dates back to the arguments of Ancient Greek philosophers about whether matter is infinitely divisible. Democritus led those who believed that eventually the fundamental atomic components of matter would be reached and further subdivision would then be impossible. The word *atom* simply means indivisible. We now know that atoms are not really indivisible, but they form an extremely important level of structure in the universe, so the ancient atomists were at least partially correct. Democritus (Figure 30) and his followers had little evidence to back up their views, which were based on rational deduction rather than experimental observation, so their insight is rather remarkable.

The modern understanding of atoms began towards the end of the nineteenth century with the discovery of the particle that we know as the electron. In the following decade, Ernest Rutherford led a team in Manchester who discovered the nucleus of the atom. Rutherford was first to realize that the nucleus contains

Figure 30 Stamp commemorating the opening of the Democritus Nuclear Search Centre.

The Cosmic Mystery Tour. Nicholas Mee, Oxford University Press (2019). © Nicholas Mee.
DOI: 10.1093/oso/9780198831860.001.0001

most of the mass of an atom, and this is surrounded by a swarm of electrons.

But just how small is an atom?

Visiting the Nanoworld

On average the distance between the electron and the nucleus in a hydrogen atom is one-twentieth of a nanometre, so the diameter of a hydrogen atom is about one-tenth of a nanometre. Carbon atoms are about twice this diameter and even the biggest atoms, such as gold, are not much larger. Gold atoms are about three times the diameter of a hydrogen atom. But what does this mean in everyday terms?

A Trillion Trillion Atoms in the Palm of One Hand

Ten million nanometres is one centimetre, so one hundred million carbon atoms in a line would stretch just two centimetres. In a cube with edges two centimetres long we could pack one hundred million times one hundred million times one hundred million carbon atoms—that is one trillion trillion atoms, or one followed by twenty-four zeros: 1,000,000,000,000,000,000,000,000. Whether it be a priceless diamond or just a humble piece of graphite, we could easily hold a cubic block this size between our fingers. Yet it is composed of around one trillion trillion carbon atoms! Atoms do not vary that much in size, so a handful of any solid material contains somewhere in the region of a trillion trillion atoms, give or take the odd gazillion.

The World's Sharpest Needle

It was once said that seeing pictures of atoms would always be impossible. Gerd Binnig and Heinrich Rohrer changed all that. In 1981, while working at IBM Zurich, they invented the scanning tunnelling microscope (STM), which maps the contours of the electric fields within atoms down to a resolution of one-hundredth of a nanometre. The STM is constructed around an extremely fine needle with a tungsten or gold tip that tapers down to a single atom. As the needle scans across a sample, an electronic feedback mechanism controls the position of the needle tip with astonishing precision. The needle scans the surface at different heights to within a fraction of an atomic diameter and these measurements are fed into a computer, where three-dimensional rendering software generates an image of the surface. The resulting pictures are remarkable, showing the positions of individual atoms.

Entering a New World in a Grain of Sand

The STM can even pick up and arrange atoms one by one. Forty-eight atoms were positioned individually using an STM to create Figure 31. It shows the electric fields of a ring of iron atoms placed on a wafer of copper. Each atom appears as a cyan-coloured peak.

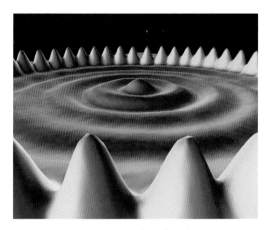

Figure 31 *Quantum Corral*, image produced with a scanning tunnelling microscope.

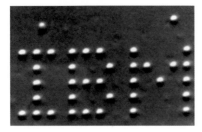

Figure 32 The IBM logo written in xenon atoms by Don Eigler.

More amazing still is the distribution of electron waves that can be seen within the *corral* of atoms.

The IBM researchers were stunned by the beautiful pictures they had created. 'I could not stop looking at the images,' said Binnig. 'It was like entering a new world.' Binnig and Rohrer were rewarded for opening up this new world with the 1986 Nobel Prize in Physics.

Don Eigler used the mind-boggling capabilities of the STM in 1989 to create the world's smallest advert, spelling the name of his employers with thirty-five xenon atoms positioned on the surface of a nickel crystal (Figure 32).

IBM researchers even used the STM to create the frames for an atomic scale animation called *A Boy and his Atom: The World's Smallest Movie* (Figure 33). It is certainly worth taking a look at this video on

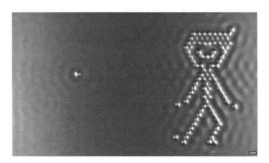

Figure 33 Frame from *A Boy and his Atom* by IBM.

YouTube. It really is quite a mind-bending achievement to create an animated character from the manipulation of individual atoms.

Onwards and Inwards

Atoms may be small, but the nucleus is minuscule. Compared to the size of an atom, it is like an apple seed in a large concert hall such as London's Royal Albert Hall. Rutherford and his teams, first in Manchester and later in Cambridge, delved even further into atomic structure. Their experiments were performed using the radioactivity of one substance to disrupt the nuclear structure of another. Radioactive elements such as uranium, radium and polonium emit high energy alpha particles and it was these projectiles that enabled Rutherford's team to investigate the composition of atoms.

Rutherford identified the first subcomponent of the nucleus in 1917. It is a positively charged particle known as the proton, with almost 2000 times the mass of the electron. Rutherford

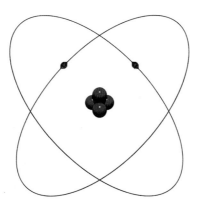

Figure 34 The nucleus of a helium-4 atom contains two protons and two neutrons. The nucleus is orbited by two electrons. (The size of the nucleus is greatly exaggerated.)

suspected as early as 1920 that the proton might have a neutral sister particle—the neutron—of similar mass. His team was on the lookout during the next decade, but it was not until 1932 that Rutherford's colleague James Chadwick finally tracked it down. The clear implication of this breakthrough was that atomic nuclei are composed of tightly bound collections of two kinds of particle: protons and neutrons. For instance, alpha particles are actually identical to helium nuclei and are formed of two protons and two neutrons bound together (Figure 34), whereas a much larger gold nucleus is composed of 79 protons and 118 neutrons. The discovery of the neutron triggered the start of the nuclear age and all the ominous responsibilities that come with our deepening knowledge of the structure of matter (Figure 35). Within a decade the first nuclear reactors were constructed and just a few years later nuclear weapons were detonated.

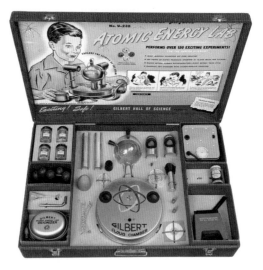

Figure 35 A. C. Gilbert Company's U-238 Atomic Energy Lab from the 1950s.

Deeper and Deeper

To probe the structure of matter further it was necessary to raise the energy of the projectiles and this required the construction of particle accelerators or atom smashers, as they were known. In these machines an electric field is used to accelerate charged particles, such as electrons or protons, which are then blasted into a target material. The colliding particles are scattered or transformed into new particles and the resulting particle debris is then analysed for any interesting events that may have occurred. Many new particles have been discovered in this way since the 1950s. This research continues on a much grander scale than ever before with the highest energy accelerator of all time, the Large Hadron Collider. It is now possible to peer inside protons and neutrons and we find an even deeper layer of structure buried within. Protons and neutrons are composed of particles known as quarks. All will be revealed in a later chapter.

Twinkle, Twinkle Little Star

Twinkle, twinkle little star
How I wonder what you are!

Robert Bunsen and Gustav Kirchhoff opened up a whole new world of science with a sensational discovery in 1859. Bunsen excitedly announced to a colleague:

> At present Kirchhoff and I are engaged in a common work which doesn't let us sleep.... Kirchhoff has made a wonderful, entirely unexpected discovery in finding the cause of the dark lines in the solar spectrum.... Thus a means has been found to determine the composition of the Sun and fixed stars with the same accuracy as we determine sulphuric acid, chlorine etc, with our chemical reagents. Substances on the Earth can be determined by this method just as easily as on the Sun, so that, for example, I have been able to detect lithium in twenty grammes of sea water.

The Chemical Fingerprint

Bunsen and Kirchhoff had discovered that each chemical element has a unique fingerprint, allowing them to track it down anywhere in the universe (Figure 36). If an element is heated in a flame until it glows, it emits light of a characteristic colour. And when this light is passed through a prism, the resulting spectrum appears as a number of bright lines corresponding to a series of sharply defined wavelengths. For instance, sodium will burn with

The Cosmic Mystery Tour. Nicholas Mee, Oxford University Press (2019). © Nicholas Mee.
DOI: 10.1093/oso/9780198831860.001.0001

Figure 36 A few of the stars at the heart of the globular cluster Omega Centauri.

an intense yellow glow. Many street lamps contain sodium vapour and this is why they shine with a yellow hue. When passed through a prism, sodium light splits into a pair of sharp lines in the yellow region of the spectrum. These two lines reveal its presence anywhere in the universe—this is the fingerprint of sodium. Bunsen and Kirchhoff took each element in turn and heated it in a flame until it glowed. They then passed this light through a prism and recorded its unique fingerprint. You may remember the burner developed by Bunsen for this work from your school chemistry lessons.

Cosmic Barcode

Light from a hot body, such as a star, is generated by the random jostling of particles in its outer layers. This light is emitted at all wavelengths, so when dispersed with a prism it forms a continuous band or spectrum. If an element is present in the atmosphere of a star it will absorb some of the light streaming out of the star, but only at certain wavelengths. The star's light is therefore depleted at these wavelengths and they appear as dark lines in the star's spectrum. Bunsen and Kirchhoff discovered these dark lines are at exactly the same wavelengths as the bright lines that form that element's spectrum in the laboratory. This gives a straight-forward method for detecting the element in the star's atmosphere. It would be over half a century before anyone could explain how atoms interact with light, so it was not clear why the technique worked, but all the same it worked like a charm.

A star's spectrum is like a cosmic barcode containing precise information about the identity of each element it contains. By matching the dark lines in the spectra of the Sun or any other star to the lines they were observing in the laboratory, Bunsen and Kirchhoff could recognize the elements present in the star. The telltale signs of each element are clear even at a distance of trillions of kilometres. Figure 37 shows the solar spectrum above the emission spectra of four elements. The dark lines in the solar

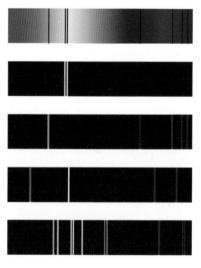

Figure 37 From top to bottom: spectrum of the Sun, emission spectra of sodium, hydrogen, lithium and mercury.

spectrum match the bright lines in the spectra of sodium and hydrogen, but not lithium or mercury, clearly indicating that the solar atmosphere contains sodium and hydrogen, but not lithium or mercury.

How amazing it must have been for Bunsen and Kirchhoff when they first determined the composition of the stars.

The Mighty Atom

In 1913 Niels Bohr made one of the great breakthroughs in our understanding of the world. He constructed a model of an atom in which the electrons orbit the nucleus in a fixed set of energy levels. We now know this works because electrons behave like waves and only waves of certain precise wavelengths will wrap around the atom without overlapping themselves. Bohr realized his model could explain the lines in the spectrum of an atom. When an electron falls from one energy level to a lower energy level it emits a particle of light, or photon, whose energy is equal to the difference in energy between the two levels. (This is what is represented by the equation shown on the stamp in Figure 38.) As the electron can only occupy a discrete set of energy levels, the photons emitted in this way only have a certain set of energies, each corresponding to a different colour. So the spectrum of any material formed from these atoms has its own characteristic set of spectral lines, just as Bunsen and Kirchhoff had observed.

Bohr's model was a great success and won him a Nobel Prize in 1922. However, Bohr felt his model should be capable of more. If it really represented the structure of an atom, then it should explain how atoms bond in chemical reactions and why each atom has its own particular chemical properties. Bohr made some progress in answering these questions, but eventually had to admit defeat. He passed the problem over to a young Austrian prodigy named Wolfgang Pauli, who had made a name for himself by writing a textbook on the new subject of relativity at the age of just 19.

Figure 38 Stamp issued in Denmark to commemorate the fiftieth anniversary of the invention of the Bohr model of the atom by Danish physicist Niels Bohr.

Don't Park Your Electrons in My Space

In 1925 Pauli came up with the fix that would enable Bohr's model to *explain* chemistry. The electric charges of protons and electrons are equal in size, but protons are positively charged, whereas electrons are negatively charged. So in an electrically neutral atom the number of orbiting electrons equals the number of protons in the nucleus. Pauli knew that in Bohr's original model all the electrons would fall to the lowest energy level of the atom and just sit there. If this were true, it is hard to see how the chemical properties of atoms could be explained. Pauli realized that this puzzle could only be solved if electrons behave in a surprising way. His explanation is known as the exclusion principle. It states that each quantum state is only available to a single electron. This means that, like a parking space, a quantum state is either vacant or occupied by one electron at any moment in time.

The exclusion principle restricts the number of electrons that can have a particular energy to the number of available slots at that energy. Chemists refer to the collection of available slots at each energy level as a shell. The equations of quantum mechanics determine the number of slots in each shell and it goes as follows: 2, 8, 8, 18.... In a hydrogen atom there is a single electron and it falls into the lowest energy level. In helium, which has two electrons, both electrons fall to the lowest energy level. The two slots in the first shell are now filled. Lithium atoms have three electrons, two occupy the lowest energy level and fill the first shell, but the third electron can only fall as far as the second energy level, so it becomes the first electron in the second shell. The arrangement of electrons in these and subsequent atoms is shown in Figure 39. Electrons fill the shells step by step, with no more than one electron occupying each available slot.

Pauli realized that only the outermost electrons—those in the partially filled highest energy shell—take part in chemical reactions. It is these outer electrons that account for the chemical properties of an element. Elements with the same number of electrons in the outermost partially filled shell of their atoms have similar chemical properties and this explains why the Periodic

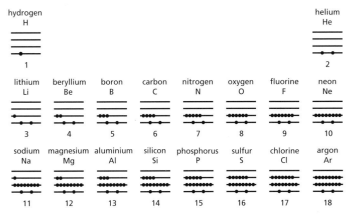

Figure 39 The figure shows a schematic representation of the energy levels in atoms. The number of available slots in each shell is 2, 8, 8, 18. Electrons fall into the lowest available slot and only one electron can occupy each slot. The number of electrons in each atom is determined by the number of protons in its nucleus.

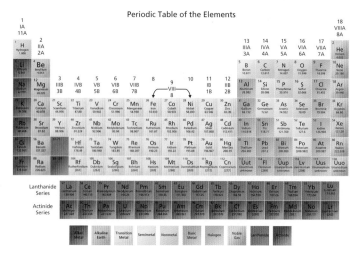

Figure 40 The Periodic Table of the Chemical Elements.

Table of the Elements captures so much that is important in chemistry. For instance, sodium (Na) and potassium (K) are chemically very similar. This is because they both have one electron in their outermost shell (the third shell in the case of sodium and the fourth shell in the case of potassium) and this is why they both appear in the first column of the Periodic Table (Figure 40). Pauli was awarded the Nobel Prize in Physics in 1945 for this work.

Spin and Statistics!

Fundamental particles come in two classes distinguished by their collective behaviour. Matter is formed from particles that obey the exclusion principle. These include electrons, protons, neutrons and quarks. Those in the other class do not obey the exclusion principle. These include photons and their relatives that

Figure 41 Wolfgang Pauli.

generate forces when passed between other particles. The class to which each type of particle belongs is determined by a rather surprising feature of the particle—the rate at which it spins! Elucidating this important property of particles was among the many great contributions that Pauli made to fundamental physics (Figure 41). We will take a closer look at particles and their interactions in the next chapter.

Forces of the World Unite!

The alchemists' dream was to understand and control the structure of matter and to turn base metal into gold. Modern physicists are equally curious, but hopefully rather less avaricious.

Four Forces

The twentieth century saw the development of machines to delve deep into the structure of matter. John Cockcroft and Ernest Walton constructed the first particle accelerator that could interrogate the atomic nucleus in 1932. It is now in the Science Museum in London and looks like a machine that Dr Frankenstein would have been proud of (Figure 42).

By the middle of the century all known phenomena could be described in terms of just four forces. Gravity controls the large-scale structure of the universe, but the intrinsic weakness of gravity means its effect is utterly negligible between pairs of fundamental particles, such as electrons in atoms. By contrast, the other three forces all play important roles in particle physics. The electromagnetic force holds atoms together. It gives us the electrical industry and the whole science of chemistry. The strong force binds the nucleus of an atom together. The weak force is responsible for some types of radioactivity and plays a critical role in synthesizing atoms in the stars.

The Cosmic Mystery Tour. Nicholas Mee, Oxford University Press (2019). © Nicholas Mee.
DOI: 10.1093/oso/9780198831860.001.0001

Figure 42 The Cockcroft–Walton machine.

A Modern Alchemist

There has been an equally impressive consolidation in understanding the particles on which these few forces act. During the 1950s and 1960s particle accelerators revealed new particles at a disconcerting rate. Murray Gell-Mann was the leading architect of new ideas that would bring order to the particle mayhem. Known for his broad spectrum of interests, which range from linguistics to ornithology, he is shown in Figure 43 playing an arithmetical African game known as oware. The game is said to have a

Figure 43 Gell-Mann playing oware.

Figure 44 Left: proton; right: neutron, where d labels the down quarks and u labels the up quarks.

symbolic significance. The board is sometimes set east to west to align with the rising and setting Sun. If the board is the world, the stones are the stars and the cups are the months of the year. Moving the stones is said to mimic the gods moving through space and time.

Gell-Mann discerned a deeper substructure to the multitude of particles that feel the strong force. He proposed in 1964 that these particles, including protons, neutrons and their more exotic relatives, are formed from a new class of fundamental particles called *quarks*—a nonsense word Gell-Mann borrowed from a line in James Joyce's surreal wordfest *Finnegans Wake*.

Initially Gell-Mann deduced the existence of three types of quark known as *up*, *down* and *strange*. Three more quarks have since been discovered: the *charm*, *bottom* and *top* quarks, making a total of six. A similar model was proposed by George Zweig, who referred to these subcomponents as *aces*. A key feature of these models is that protons contain two up quarks and one down quark, while neutrons are formed of two down quarks and one up quark (Figure 44). Particles composed of quarks are known collectively as *hadrons*.

QED

In the late 1940s Richard Feynman, Julian Schwinger, Sin-Itiro Tomonaga and Freeman Dyson devised a quantum theory of electromagnetism. Quantum electrodynamics, or QED as it is usually known, provides an incredibly precise explanation of the electromagnetic force at the level of particle interactions. According to QED, electrons, protons and other charged particles interact by passing photons—the fundamental particles of light—back and forth. Feynman is celebrated on the stamp shown in Figure 45, where he is depicted along with the diagrams

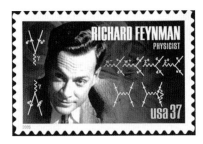

Figure 45 Richard Feynman with some of his particle interaction diagrams.

he invented to represent particle interactions. QED has also proved to be the perfect model for constructing theories of the other forces.

Gluing the Quarks Together

The strong force is described by a theory called quantum chromodynamics (QCD) in which quarks interact through the exchange of particles known as *gluons*. Why are they called gluons? Because they provide the *glue* to hold quarks together, of course. QCD is like QED on steroids. It is the same sort of theory, but while there is just one photon, there are eight types of gluon. Also, photons are uncharged and therefore don't interact with other photons, whereas gluons feel the strong force, so they do interact with other gluons. This makes QCD calculations more complicated, but the theory works incredibly well and provides a wonderful description of the workings of nature at very short distances, such as within the nucleus of an atom.

Unification

The weak force has its own idiosyncrasies. It operates through the exchange of three particles known as the W^+ (W-plus), W^- (W-minus) and Z bosons. Unlike photons, which are massless, this trio are heavyweights. It is as though the weak force operates by lobbing cannonballs between particles, rather than peas, which accounts for its weakness and short range. Peter Higgs and two other teams of physicists independently invented a new quantum field to give mass to the weak force exchange particles. If the theory were correct, the electromagnetic and weak forces would now be seen as two aspects of a single unified *electroweak force*, despite their obvious apparent differences. A critical prediction of the theory is the existence of a new fundamental particle—the Higgs boson.

The Large Hadron Collider (LHC) is the modern descendant of early atom smashers such as the Cockcroft–Walton machine. It is the highest energy accelerator of all time (Figure 46). In the LHC the projectiles are protons, which are accelerated in two beams that intersect at four points around the machine where head-on collisions occur. Hence the name—it is a large machine, 27 kilometres in circumference, that smashes protons, which are hadrons, head-on into each other. Almost all the particles created in these collider experiments are unstable, they race away from the impact point and rapidly transform into two or more

Figure 46 The beam pipe of the LHC during construction.

lighter particles that carry away the released energy. Eventually the only particles that remain are members of a small collection of stable particles. These include protons, electrons, photons and neutrinos. If these particles were not stable there would be no atoms and no light.

On 4 July 2012 CERN announced the LHC had found the Higgs boson. This completed the unification of the electromagnetic and weak forces—the greatest unification of forces since Faraday and Maxwell 150 years ago.

A Triumph of Modern Physics

The combination of the electroweak force and the strong force is known as the *Standard Model*. This incredibly successful theory is one of the great triumphs of modern physics. It explains the structure of matter in terms of a handful of fundamental particles and a couple of forces.

Fundamental particles divide naturally into two classes: fermions and bosons. Fermions are the particles of which matter is composed. They obey the exclusion principle and are named after Enrico Fermi. Bosons are particles such as photons that are exchanged between other particles to produce forces. They are named after Satyendra Nath Bose.

There are just twelve fundamental fermions along with their antiparticles. These fermions form three *generations* of four particles. The first generation consists of the *up* and *down* quarks, the *electron* and the *electron neutrino* shown in the first column of Figure 47. Ordinary matter is formed from the first three of these particles. Each particle in the second and third generations carries the same charges and seems to be just a heavier replica of the corresponding particle in the first generation. These particles are shown in the next two columns.

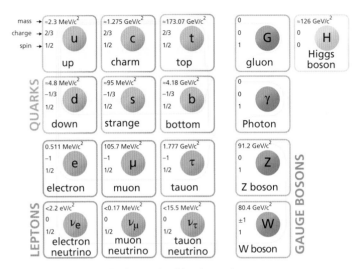

Figure 47 The Standard Model Table of Particles.

The right-hand part of the table shows the fundamental bosons, responsible for producing the strong, electromagnetic and weak forces. The discovery of the Higgs boson filled the last empty place in the table.

It is a remarkable fact that the Standard Model is consistent with every particle physics experiment ever performed. For instance, LHC researchers released data in 2016 of how an obscure particle known as B_s^0, composed of a strange quark and a bottom anti-quark, decays into a pair of muons. The Standard Model predicts this is an extremely rare event occurring just three times in every billion decays of the particle, and this matches exactly what the researchers observed, making it the rarest decay process ever measured. It highlights the precision with which particle physics predictions are being tested, and, so far, the Standard Model agrees with the experimental results every time.

Where Do We Go from Here?

Despite the incredible success of the Standard Model it cannot be the end of the story. The Standard Model provides a very concise account of the structure of matter, but it has various loose ends and unexplained features. There is no explanation of why there are three generations of matter particles. There is no explanation of why the masses of the fundamental particles vary so greatly, as illustrated in Figure 48.

There is also the question of further unification. The Standard Model comes close, but it does not unify the electroweak force with the strong force. And although there are not many fundamental

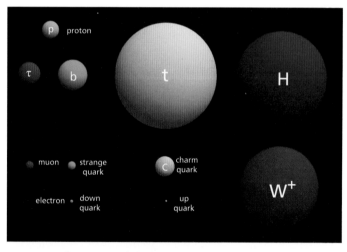

Figure 48 The masses of the fundamental particles vary greatly. The volume of each sphere shown here is proportional to the mass of the corresponding particle. The masses of the neutrinos are far too small to be visible.

particles, the number is far more than we might expect from a truly fundamental theory.

Most serious, perhaps, is the rather more disturbing issue that about 85% of the matter in the universe exists in some mysterious form that does not fit into the Standard Model Table of Particles. It has been dubbed dark matter.

Most of the Universe is Missing!

It is a rather embarrassing fact that most of the universe seems to be missing!

There are two ways to determine the amount of material in the universe. One is to measure its gravitational pull, the other is to measure the amount of light emitted by luminous objects. All objects have a gravitational attraction, but not all objects emit light (Figure 49), so we would expect the first measure to give an answer that is bigger than the second, and it does.

Figure 49 The Moon does not emit light, but it certainly has a gravitational attraction that raises the tides twice a day.

Dark Matter

Before these calculations were possible, astronomers assumed that most mass is in the form of stars, which emit lots of light, so

The Cosmic Mystery Tour. Nicholas Mee, Oxford University Press (2019). © Nicholas Mee. DOI: 10.1093/oso/9780198831860.001.0001

Figure 50 Neptune's existence was originally deduced from its gravitational influence on the planet Uranus. Only later was it observed through a telescope.

the assumption was that most of the matter in the universe would be visible. It turns out, however, that most of the matter in the universe does not emit light (Figure 50). Astronomers call the invisible stuff *dark matter* simply because we cannot see it. Fritz Zwicky was the first to draw attention to this conundrum in the 1930s. There is more about this pioneering astronomer in a later chapter, 'The Crab and the Jellyfish', in Part II.

Could There Be Some Mistake?

The first question is: could the measurements by wrong? After all, it can't be that easy to measure the properties of galaxies that are millions of light years away. But astronomers have been working on this problem for many decades and the evidence is now quite conclusive. Vera Rubin used a clever method for determining the mass of galaxies in the 1970s. She calculated the speed of rotation of spiral galaxies that we view edge-on (Figure 51) by measuring the Doppler shift of the light from their two extremities. The starlight from one side is red shifted as the stars move away from us, while the starlight from the other side is blue shifted as they

Figure 51 The Sombrero Galaxy, a beautiful spiral galaxy that we view almost edge-on.

move towards us. The rate of rotation can then be used to deduce the total gravitational mass of the galaxy. These measurements show that galaxies rotate so quickly they would fly apart if they were solely composed of the visible material they contain, implying that the galactic discs are embedded within spherical haloes of dark matter.

Gravitational lenses, such as the one shown in Figure 5, provide another means of measuring the mass of distant galaxies and galaxy clusters. They show that galaxy clusters bend light much more than expected from the amount of matter we can see. So there is little doubt the cosmos contains lots of dark matter.

The Big Mystery

The big mystery is: what is it? And this is where everything becomes much more murky, because all the obvious answers, such as dark gas clouds that have not yet formed stars or burnt-out stellar remnants cannot be the answer. The Big Bang model of the early universe works very well based on the known quantities of ordinary matter. In particular, predictions of the amount of heavy hydrogen and helium synthesized in the immediate aftermath of the Big Bang match observations. However, this agreement would be lost if the density of ordinary matter in the early universe was much greater than has been assumed.

MACHOs and WIMPs

The possible sources of dark matter fall into two categories referred to as massive compact halo objects (MaCHOs) and weakly interacting massive particles (WIMPs). MaCHOs are burnt-out stars, such as white dwarfs and neutron stars, in the halo of our galaxy. Evidence for these dark stellar remnants has been sought in surveys of the galactic halo. We cannot see them directly, but on rare occasions when such a body passes directly between us and a more distant star it produces a sudden brightening of the distant star due to gravitational lensing. These micro-lensing events have been seen, but in far smaller numbers than would be necessary to account for dark matter, so most physicists believe that dark matter consists of vast quantities of WIMPs. A WIMP is a stable particle that only interacts very weakly with ordinary matter, probably only gravitationally and certainly not electromagnetically. This would explain why dark matter forms a separate component of the material around galaxies and also why WIMPs have not already turned up in particle collider experiments.

Cosmic Relics

Neutrinos were created in vast quantities in the early universe and are still being produced by stars and supernovae. It was once thought this might explain the origin of dark matter. The mass of the neutrinos is too small, however, to account for much of the dark matter, even in the gargantuan quantities that we know are out there. Most physicists are convinced that dark matter is composed of a new unknown type of stable particle that was produced in large quantities in the very early universe. Finding such a particle is one of the main targets of research at the Large Hadron Collider.

Detector Experiments

The LHC experimenters are hoping to track down newly created dark matter particles in the debris of their proton collisions. But there is another way to find these elusive particles. If the universe is full of the stuff, there should be plenty of it about. A host of laboratories around the world are waiting for the dark matter particles to come to us. One such is the XENON100 experiment buried deep underground in the Gran Sasso laboratory in central Italy (Figure 52). The underground location shields the detectors from false alarms caused by cosmic rays and other background

Figure 52 The XENON100 experiment in Gran Sasso, Italy.

particles. The experiment uses 165 kilograms of liquid xenon held in containers made from materials with ultra-low background radioactivity. The liquid xenon is monitored for the occasional collision between an unidentified intruder particle and an electron in a xenon atom. Each such collision ionizes the xenon atom and sends the electron racing through the fluid, emitting ultraviolet photons, producing an identifiable signal that an interaction has occurred.

A wide variety of experiments have been in operation around the world for several years in the quest for dark matter. So far none have found any sign of this mysterious substance. But it took 48 years to find the Higgs boson, so no one is giving up yet. The researchers will endeavour to increase the sensitivity of their detectors and will continue looking in the hope that they might be first to bag a WIMP.

Other Explanations

But perhaps there is another explanation. Some people certainly think so. It has been suggested there is a good reason why no one has found a WIMP and that is because they don't exist. No one is questioning the evidence for the super-sized gravitational attraction of galaxies and galaxy clusters, but there are some who question its interpretation. Their conclusion is that we need a new theory of gravity. They claim that no dark matter is required if gravity is stronger than expected when acting over very long galactic-scale distances. This remains a marginal view, but such theories have been proposed. None of them offer a compelling alternative to general relativity, however. General relativity is widely acknowledged as the most elegant theory ever devised and most physicists would find it hard to accept an alternative unless it was equally beautiful. Furthermore, the predictions of general relativity have been verified to great accuracy and its remarkable consequences continue to be confirmed. It is a tall order for a new theory to match all the successful predictions of general relativity and do away with dark matter as well.

Biting the Bullet

The *Bullet Cluster* is a distant collection of galaxies 3.7 billion light-years away that formed through the collision of two galaxy clusters (Figure 53). It is invoked as an astronomical object whose structure can only be explained by the presence of large quantities of dark matter and not by any new theory of gravity. This is because the distribution of dark matter in the cluster does not match the distribution of the visible matter.

The cluster contains two ultra-hot gas clouds containing most of its ordinary matter—far more than is contained in all the

Figure 53 The Bullet Cluster. In this false-colour image the distribution of ordinary matter is shown in red and the distribution of dark matter is shown in blue.

galaxies. The collision of the component clusters has heated the gas so much that it emits X-rays and this is how it is detected. The gas does not emit visible light, so it is given a false red colouring in the image. The distribution of dark matter in the cluster was mapped by analysing the gravitational lensing of light from background galaxies. Its presence is indicated with a false blue colouring in the image.

The mass of the dark matter is far greater than the combined mass of the ultra-hot gas clouds and the visible matter in the galaxies. Even more significant is the separation of the dark matter and the ordinary matter. In the orthodox gravity plus dark matter theory this is no surprise, as we would expect that prior to and during the galaxy cluster collision, the distribution of hot gas would be affected by both electromagnetic and gravitational interactions, whereas the dark matter would only be affected by gravity. By contrast, it is hard to reconcile these observations with alternative gravity theories in which the gravitational lensing around the cluster is explained by enhancing the attraction of ordinary matter when acting over extremely long distances, as the ordinary matter is in the wrong place to account for the observed lensing. The Bullet Cluster is therefore regarded as direct evidence that dark matter must exist.

The identity of dark matter remains a mystery. But one that physicists expect to solve in the not too distant future.

PART II

THE HISTORY, GEOGRAPHY AND ARCHITECTURE OF THE COSMOS

From Genesis to Revelation

Modern cosmology is just one hundred years old. Poets, seers and sages have, of course, contemplated the origin of the universe for millennia and arrived at various conclusions, but there aren't really that many distinct possibilities. The universe is finite, eternal or cyclic, and the third of these possibilities is like a combination of the previous two, an infinite sequence of finite cycles.

The finite cosmos of the Judaeo-Christian tradition is an example of the first option (Figure 54). It opens with:

In the beginning God created
the heaven and the earth.

and ends with the Apocalypse. Writing in the fifth century, St Augustine claimed the only logical conclusion was that time began with the Creation.

The Island Universe

The second option corresponds to the eternal cosmos of Aristotle, who believed the universe to be a sphere surrounded by the Void. Aristotle's cosmos was enclosed by the sphere of fixed stars whose distance was not much greater than that of Saturn. Aristotle held the view that although space is finite, it has always existed, so time is infinite.

The Cosmic Mystery Tour. Nicholas Mee, Oxford University Press (2019). © Nicholas Mee.
DOI: 10.1093/oso/9780198831860.001.0001

Figure 54 *The Ancient of Days* by William Blake.

The Ring Cycle

According to Hindu cosmology the universe is eternal and cyclic. There is an infinite sequence of universes that each emerge from chaos, to grow and flourish, then decay, die and dissolve into the chaos from which the subsequent universe is born. Each cycle takes an immense period of time, the *kalpa*, which lasts upwards of 4,320 million years.

There are parallels with Nordic cosmology where each world cycle ends with Ragnarok or, if you are an opera lover, Götterdämmerung—Twilight of the Gods (Figure 55).

Figure 55 Götterdämmerung.

Einstein's Static Universe

Einstein completed his general theory of relativity in 1915. Formulating this new theory of gravity was possibly the greatest intellectual achievement ever made by a single individual. By 1917 he was ready to apply his new theory to the entire cosmos. But even Einstein had preconceived ideas about how the results should turn out. His initial calculations seemed to imply the universe must be either expanding or contracting, which clashed with

Einstein's expectations that the universe was eternal. Einstein realized, however, that he could add an extra term to the equations—the cosmological term—that would only alter his theory when applied over huge cosmological distances. The inclusion of this term allowed a solution that was neither contracting nor expanding, but balanced on a knife edge between the two. This was Einstein's static universe. It was an island universe in some ways similar to Aristotle's.

The Cosmic Egg

Georges Lemaître, a priest from Belgium who became enchanted with general relativity while studying theology in Cambridge, took a different view (Figure 56). He published his results in 1927 and concluded that Einstein's theory implies the universe is expanding and therefore must have begun at some point in the

Figure 56 Georges Lemaître.
© BPost

Figure 57 The Cosmic Egg.

distant past. Lemaître dubbed this cosmic origin the *primaeval atom* or, even more poetically, the *cosmic egg* (Figure 57).

Einstein was not enthusiastic about Lemaître's conclusions and pointed out that a Russian physicist Alexander Friedmann had derived similar results five years earlier. He told Lemaître: 'Your calculations are correct, but your grasp of physics is abominable.'

In physics, authority counts for little, however. Ultimately there is one arbiter who decides the validity of a theory—the real world. Lemaître realized there is a consequence of an expanding universe that could be tested. In an expanding universe the further away any light source, the faster it will be receding, and the faster it is receding, the greater the red shift of light in its spectrum.

Red Shift

Light from a hot body, such as a star, is generated by the random jostling of particles in its outer layers. This light is emitted at all wavelengths, so when dispersed with a prism it forms a continuous band or spectrum. Galaxies contain lots of stars and lots of clouds of hydrogen and other gases. The electrons in the gas atoms absorb light of specific wavelengths that promote them to higher energy levels in the atoms. This depletes the light we receive from a galaxy at precisely these wavelengths, forming dark *absorption lines* in the spectrum, as shown in Figure 58. (The electrons later emit light of exactly the same wavelength, but this is emitted in all directions, so only a tiny fraction is directed towards us.)

If the galaxy is receding from us, the wavelength of the absorption lines shifts. This is analogous to the fall in pitch of sound waves from a siren when an ambulance or fire engine moves away from us. When applied to light, it is known as red shift, because red is the long wavelength end of the visible spectrum, even though the lines could be in any part of the spectrum, such as infrared, ultraviolet or radio waves. Similarly, light from an approaching source is blueshifted.

Lemaître showed that in a steadily expanding universe the more distant a galaxy, the faster it is receding from us, which

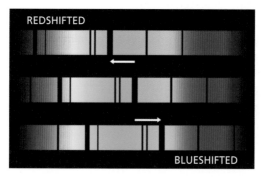

Figure 58 Schematic illustration showing (centre) absorption lines as they would appear in the laboratory, (top) redshifted spectrum from a receding source, (bottom) blueshifted spectrum from an approaching source.
Credit: Christopher S. Baird

implies that the red shift of a galaxy's light is proportional to the distance to the galaxy. There was just one problem with using red shift to determine whether the universe is expanding. How do we know the distance to each galaxy?

The Distance to the Stars

Henrietta Leavitt (Figure 59) was employed as a *computer* at Harvard College Observatory. She was given the task of studying and cataloguing variable stars, a topic centuries old that must have seemed an unlikely subject for a major discovery. However, Leavitt found a remarkable pattern in a class of stars known as Cepheid variables. These are very bright giant stars that can be seen from immense distances. They go through regular cycles of brightening and dimming.

By studying a collection of Cepheids all lying at essentially the same distance in the Large Magellanic Cloud, a dwarf galaxy

Figure 59 Henrietta Swan Leavitt (1868–1921).

orbiting the Milky Way, Leavitt found that the brighter the Cepheid, the longer its period. So by measuring the period of a Cepheid we can deduce its true luminosity and by comparing this to how bright it appears in our skies we can determine how far away it is. Her original study was published in 1908 and this was followed by a more detailed account in 1912. Tragically, Leavitt died in 1921 before anyone realized the significance of her discovery.

Cepheid variables are named after the star Delta Cephei whose variability was discovered by John Goodricke in 1784. There is more about this remarkable astronomer in a chapter in Part III: 'The Gorgon's Head'.

You may have seen a Cepheid variable; the closest such star is very well known. It is the pole star, Polaris, whose position near the celestial pole has helped travellers navigate for several centuries (Figure 60). Although Polaris is not one of the very brightest stars

Figure 60 Polaris and the Pointers. In the UK, the Big and Little Dippers are usually known as the Plough and the Little Bear.

in the night sky, this is only because it lies 433 light years distant, fifty times further away than Sirius (which is the brightest). Polaris is a yellow supergiant with over five times the mass of the Sun. Its brightness varies over a period of four days, but this is not easily noticeable to the naked eye.

Living in an Expanding Universe

Edwin Hubble tracked down Cepheid variables in a number of relatively nearby galaxies with the 100-inch Hooker Telescope at the Mount Wilson Observatory in California, and used them to calculate the distance to each galaxy. He compared this data to the spectra of these galaxies compiled by Vesto Slipher and Milton Humason. Hubble announced in 1929 that the more distant a galaxy, the faster it is racing away and concluded that we live in an expanding universe.

Lemaître was proved correct, but he has been sidelined somewhat in the history of cosmology; perhaps because physicists were uncomfortable that their leading theory for the origin of the universe was proposed by a Catholic priest. Lemaître is remembered more for Einstein's criticisms than his actual work, which is rather unfair, as in this case Einstein was wrong. Nevertheless, Einstein and Lemaître became firm friends.

Not everyone was convinced that the universe emerged from nothing in the distant past. The battle for the cosmos would continue for several decades. In the next chapter we will see how it played out.

The Battle for the Cosmos!

Following the Second World War, a battle for the origin of the universe was waged by two teams of nuclear physicists, one led by George Gamow, the other by Fred Hoyle. Both would be incredibly successful, but not quite in the ways they expected.

Gamow was a brilliant physicist from Russia who gained international acclaim at a young age for using the new quantum theory to understand radioactive decay. Gamow had an irrepressible sense of humour and was a great promoter of science, explaining the latest scientific ideas through the adventures of a certain Mr Tompkins (Figure 61).

Hoyle was a leading British astrophysicist who, like Gamow, was an avid science communicator, writing science fiction novels and television series, as well as presenting BBC radio programmes. Hoyle was the arch critic of the view that the universe began in the relatively recent past. Speaking on BBC Radio in 1950 he dismissed the idea of its eruption and expansion from an incredibly hot, dense fireball as simply the *Big Bang Theory*.

The Steady State Theory

Hoyle considered fundamental biochemistry to be so complex that the origin of life was almost inconceivably improbable; so improbable that even billions of years were insufficient for it to arise naturally. He believed that life could only exist in an eternal

The Cosmic Mystery Tour. Nicholas Mee, Oxford University Press (2019). © Nicholas Mee.
DOI: 10.1093/oso/9780198831860.001.0001

Figure 61 An illustration from one of Gamow's popular science classics showing Mr Tompkins travelling at relativistic speeds on his bicycle.

universe, spreading to infect the entire cosmos from one star system to the next—a notion known as *panspermia* (Figure 62).

Hoyle imagined the universe to be everlasting and spatially infinite. He thought it should look pretty much the same when viewed from any point and at any time. Along with two colleagues, Thomas Gold and Hermann Bondi, Hoyle developed a highly original model of the universe based on this assumption and published it in 1948 as the *Steady State Theory*.

The biggest challenge for the Steady Statesmen was to explain how an expanding universe might appear unchanging, as the

Figure 62 Fred Hoyle on the cover the May 1954 edition of *Newsweek*.

expansion should thin out the matter it contains. Their solution was to postulate the continuous creation of matter at a rate that would compensate for this dissipation and maintain the universe's observed average density. Most of the universe is very empty space, so the required rate of matter creation is tiny. They argued that, if we are prepared to allow the appearance of huge quantities of matter at one point in time—the Big Bang—why shy away from continuous creation of matter in quantities so tiny that it could not be detected? After all, if the laws of physics are presumed to be fixed and unchanging, it is more reasonable that matter should be created continuously rather than in a single miraculous event.

So which is correct: Big Bang or Steady State? There appeared to be one way to settle the debate. The universe contains a diverse array of different atoms. According to Gamow they must have been synthesized during the Big Bang. According to Hoyle they were forged in the stars. But who was right? The battle lines were drawn, the whole universe was at stake.

The A, B, C of Cosmology

Many physicists in the United States spent the war years developing nuclear weapons in the Manhattan Project. Gamow would have been ideally suited for a role in the project, but did not have the security clearance. Throughout the 1940s he led a team researching nuclear physics at George Washington University in the American capital.

Gamow gave the task of working out the details of how atoms were created in the Big Bang to his research student Ralph Alpher for his PhD thesis. When Alpher was preparing his dissertation for publication in 1948, Gamow insisted they must ask the nuclear physicist Hans Bethe to contribute to the paper. He explained that Bethe was such a great physicist he would certainly have important insights to add and it would be wonderful if their account of the origin of the universe was authored by Alpher, Bethe and Gamow. (Alpha, beta and gamma are the first three letters of the Greek alphabet used by physicists every day in their mathematical notation.) So the paper 'The Origin of the Chemical Elements' duly appeared with the three alphabetical authors. Unfortunately, the main thrust of the paper is incorrect. During the Big Bang, the extreme temperatures required for nuclear fusion did not last long enough for the creation of atoms beyond the lightest: hydrogen, helium and lithium.

Gamow later joked that when physicist Robert Herman joined the team, he stubbornly refused to change his name to Delter (delta being the fourth letter of the Greek alphabet).

Figure 63 Robert Herman (left) and Ralph Alpher (right) opening a bottle of ylem.

The Genie in the Bottle

Gamow and Alpher named the original material of the universe *ylem*, an obscure Middle English theological term derived from the Greek *hyle*—the primordial substance from which matter is formed in Aristotelian philosophy. The genie Gamow is emerging from a bottle of ylem in the photograph in Figure 63.

The Hot Big Bang

When not joking around, Gamow's team did some serious cosmology.

Alpher and Herman realized the Big Bang had a consequence that could be tested. If the theory was correct, the universe was originally filled with photons scattering off charged particles such as electrons, protons and helium nuclei. It would have cooled as it expanded until after a few hundred thousand years all the charged particles combined into atoms of hydrogen and helium. Then, as these gases are transparent, the photons would no longer interact with them and would continue travelling unhindered through the universe. So, if the Big Bang really happened, the universe should be full of this leftover radiation from its earliest years.

Each such photon last scattered off a charged particle when the average matter temperature was 3100 K, not long after the Big Bang, so the radiation would have the statistical makeup of radiation emitted by matter at 3100 K. (To convert from degrees Kelvin (K) to degrees Celsius or centigrade (°C) subtract 273.15.) At this temperature matter is 'white hot' as it mainly radiates visible light, so most of the photons would have wavelengths in the visible region of the spectrum. But the dramatic expansion of the universe since the era of last scattering would produce an extreme red shift of the radiation. Alpher and Herman estimated that it would now be in the microwave range with the signature of radiation emitted by matter at a temperature just five degrees above absolute zero.

Forging Atoms in the Stars

Meanwhile Gamow's intellectual adversary, Hoyle, worked out the physics of nuclear fusion in stars. Hoyle's earliest papers were written in the late 1940s and early 1950s. Like all science, nuclear physics is a collaborative enterprise, so Hoyle built on critical ideas developed by Gamow, Bethe and others. The details of the forging of atoms in stars were presented in a big review paper published in 1957 known as B^{2}FH after its authors Burbage, Burbage, Fowler and Hoyle, where Burbage and Burbage are the husband and wife team Geoffrey and Margaret Burbage, and Fowler is Willy Fowler. Although Hoyle is regarded as the main

Figure 64 Butterfly Nebula.

architect of the theory of stellar nucleosynthesis, Fowler was the only member of the B²FH collaboration to receive the Nobel Prize.

Hoyle and his collaborators were correct. Apart from a few of the very lightest elements, the atoms of the Periodic Table were created in stars and supernova explosions. The outer layers of ageing stars swell and eventually drift away into space producing planetary nebulae. Figure 64 shows a particularly beautiful example. Much of the material created in stars is released back into space where it merges into the gas clouds from which subsequent generations of stars condense.

So does this mean we live in a steady state universe? No, Hoyle had won the battle for the chemical elements, but he would ultimately lose the war for the cosmos.

The Background Noise of the Universe

In 1964, Arno Penzias and Robert Wilson built a sensitive horn antenna for Bell Labs in New Jersey (Figure 65). Their receiver was plagued with interference which they assumed was due to a fault in their equipment. They tried everything to eradicate the noise, but without success. Eventually, the Princeton astrophysicists Robert Dicke, Jim Peebles and David Wilkinson provided the explanation. Penzias and Wilson had discovered the cosmic microwave background, the relic radiation from the Big Bang predicted by Alpher and Herman.

Precision measurements of the microwave background show it is identical to electromagnetic radiation emitted by matter at a temperature of 2.7 degrees above absolute zero. This is less than one-thousandth of the temperature when it last interacted with matter particles in the early universe and somewhat lower than the estimate of Alpher and Herman because the age of the universe

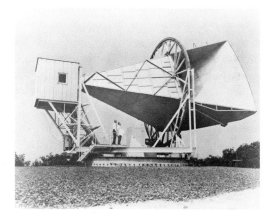

Figure 65 Penzias and Wilson with their horn antenna.

Figure 66 False colour map of the tiny variations in temperature of the cosmic microwave background across the whole sky as mapped by the probe WMAP (Wilkinson Microwave Anisotropy Probe). This is essentially a photograph of the universe 13.8 billion years ago, just 380,000 years after the Big Bang.

and other relevant cosmic parameters were not known with great accuracy at their time of writing.

The cosmic microwave background has been studied in detail by a series of space probes (Figure 66). When local effects due to our motion are taken into account, it is almost perfectly uniform across the entire sky, but there are tiny variations in temperature due to the clumping of matter in the early universe. The regions of higher density are the seeds from which super clusters of galaxies will grow. They have been mapped with great precision and are the best source of information we have about the structure and evolution of the early universe, giving us the most precise age for the universe and confirming that most of the mass in the universe is in an unknown form known as dark matter. The only explanation that accounts for the data is that the universe began 13.8 billion years ago in a hot Big Bang.

Alchemical Furnaces of the Cosmos

Arthur Eddington was raised in a Quaker family in northern England. After studying in Manchester, he won a scholarship to Cambridge University and by the 1920s he established himself as the world's leading astrophysicist. Eddington developed models of the structure and evolution of stars that form the foundations of the subject even today. The key to these models is the vast amount of thermal energy generated within a star that supports it against its tendency to collapse under gravity. But the source of all this energy was a complete mystery (Figure 67).

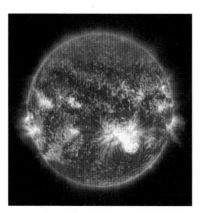

Figure 67 The Sun.

The Cosmic Mystery Tour. Nicholas Mee, Oxford University Press (2019). © Nicholas Mee.
DOI: 10.1093/oso/9780198831860.001.0001

Feeding the Stellar Furnace

Eddington was an enthusiastic advocate of Einstein's theory of relativity (Figure 68). One of Einstein's most profound insights was the equivalence of mass and energy, as expressed in his famous equation:

$$E = mc^2$$

Here E represents energy, m represents mass and c is a constant, the speed of light. Eddington realized this equation implies that converting mass into energy would provide an unprecedented power source and this might explain what keeps the stars shining. Rather prophetically, he wrote in 1920: 'If, indeed, the subatomic energy in the stars is being freely used to maintain their great furnaces, it seems to bring a little nearer to fulfillment our dream of controlling this latent power for the well-being of the human race—or for its suicide.'

Eddington had an idea for how this energy might be released in stars. Accurate new measurements showed the total mass of four atoms of the lightest element hydrogen is slightly greater

Figure 68 Einstein and Eddington.

than an atom of the second lightest element helium. Eddington reasoned that if there were some way to transform hydrogen into helium, then the difference in mass would be released as energy. This was a remarkable proposal, as nuclear physics was in its infancy, it was just ten years since the discovery of the nucleus and three years since the discovery of the proton. Nevertheless, Eddington suggested the energy of the stars might be due to the conversion of hydrogen nuclei into helium nuclei.

But there was a serious problem with the idea. Kirchhoff and Bunsen's spectroscopic methods had been used for decades to detect elements in the Sun and stars. Astronomers assumed that elements with prominent lines in the solar spectrum, such as calcium and iron, were the major constituents of the Sun and this spectral analysis appeared to show the Sun was composed of similar elements to the Earth. Indeed, the eminent astronomer Henry Norris Russell claimed that if the Earth's crust were heated to the temperature of the Sun, its spectrum would look almost the same. The composition of the Earth's crust by mass is oxygen (46%), silicon (28%), aluminium (8%), iron (6%), calcium (4%), with lesser quantities of the other elements. The proportion of hydrogen is less than 0.15%. If Russell and his colleagues were correct, then Eddington's proposal looked rather improbable, as there would be comparatively little hydrogen in the Sun. This issue would be resolved by an astronomer who was inspired by Eddington, but was severely hampered throughout her career by the institutional sexism of the time.

Undoubtedly Brilliant

Cecilia Payne won a Cambridge University scholarship in 1919 to study botany, physics and chemistry. Eddington had recently returned triumphant from an eclipse expedition to Principe in West Africa organized to test whether light bends in a gravitational field

Figure 69   Report of Eddington's eclipse expedition of 1919 in *Illustrated London News.*

as predicted by Einstein's theory of general relativity (Figure 69). The expedition found good evidence that Einstein was correct and Eddington was keen to promote his theory as the greatest scientific breakthrough since Newton. This publicity propelled Einstein towards the superstar status he would enjoy for the rest of his life. When Payne heard Eddington's captivating account of the expedition she was spellbound. She said of the lecture, 'The result was a complete transformation of my world picture. My world had been so shaken that I experienced something very like a nervous breakdown.'

Payne later nervously approached Eddington and told him she wanted to be an astronomer. He encouraged her and invited her to use Cambridge Observatory's library where she could read the latest astronomy journals. She completed her studies at the university, but was not awarded a degree, as the university did not confer degrees on women. This policy did not change until 1948. Payne realized she would be unable to pursue an academic career in England, so she left for the United States where she obtained a fellowship that would fund her graduate studies in astronomy at Harvard University.

Payne studied the spectral lines of the Sun and considered how its composition could be determined from its spectrum. She realized the assumption that the most prominent lines were produced by the commonest atoms was mistaken. Payne showed how to correctly calculate the abundance of each element from the strength of its lines and demonstrated that, although the relative abundance of elements such as carbon, silicon, iron and other heavy elements is similar in the Sun and the Earth, this is not the case for the two lightest elements, hydrogen and helium, which are vastly more abundant in the Sun. All the evidence suggested that the Sun and other stars were essentially huge balls of hydrogen and helium and the other elements were present in very small proportions. Payne submitted her doctoral dissertation describing this research in 1925. It has been described as 'undoubtedly the most brilliant PhD thesis ever written in astronomy'.

Initially, some astronomers expressed their doubts and Henry Norris Russell even persuaded Payne to include a warning suggesting there might be some error in the measurements. Within four years, however, he was convinced and the astronomy community followed his lead. Modern values for the elemental composition of the Sun by mass are hydrogen (71%), helium (27%),

Figure 70 Cecilia Payne.

with the remaining 2% shared out between all the other elements, leading with oxygen (1%), carbon (0.4%) and much smaller quantities of the rest. Payne's discovery of the true composition of the Sun and stars was fundamental to understanding the physics of the stellar furnace.

Payne made many important contributions to astronomy throughout her career, but struggled to receive the status that should have been her due. It was not until 1956 that she was appointed to a full professorship at Harvard (Figure 70).

Keeping the Lights On

We now know that Eddington was correct. The vast energy output of the Sun is produced by the gradual conversion of hydrogen into helium. The precise nuclear mechanisms by which this happens were worked out by Hans Bethe and Charles Critchfield in the late 1930s. Hydrogen fusion is the process that powers the stars for at least the first 90% of their lives, a period known as the Main Sequence.

The rate at which stars burn their fuel depends on their mass. The greater the mass of a star, the hotter its core and the faster the nuclear reactions proceed. It will take the Sun around ten billion years to burn the hydrogen fuel in its core.

Stars of less than half the Sun's mass are known as red dwarfs. A star with a mass of one-tenth of the Sun will fizzle away for 10 trillion years, which is one thousand times longer than the age of the universe. Most of the stars in the galaxy are rather feeble red dwarfs.

A star with twenty times the mass of the Sun will use up its nuclear fuel in just ten million years. Twenty times as much fuel is burnt in one-thousandth of the time, so energy is released at 20,000 times the rate of the Sun and the star will shine 20,000 times as bright as the Sun.

Cosmic Destiny

The nuclear processes that take place in the later stages of a star's life were worked out by Fred Hoyle and his colleagues. When the hydrogen nuclear fuel in the core has been converted into helium, the core contracts and if the star's mass is great enough its temperature rises until helium fusion begins. Helium is converted into carbon and oxygen in the core, and meanwhile the outer layers of the star swell up to form a bloated red giant. One such star is Betelgeuse, the shoulder of Orion the Hunter.

In five billion years' time the Earth will be engulfed by the outer layers of the Sun as it approaches the end of its life. In stars like the Sun these outer layers eventually disperse into space to reveal the star's core as an extremely dense glowing ember about the size of the Earth. The nuclear reactions in the core have now ceased and the core gradually cools as it radiates its heat into the depths of space. These stellar remnants are known as white dwarfs. They are extremely hot, but very faint as they are so small.

Figure 71 Zoom in to Betelgeuse.

The diameter of Betelgeuse is about 1000 times that of the Sun, so it would stretch as far as the orbit of Jupiter. It is so large that astronomers have even imaged its face. Betelgeuse is rather unstable and varies in luminosity quite noticeably. It is belching large amounts of gas into space, as can be seen in Figure 71.

Stellar Alchemy

In very massive stars, like Betelgeuse, further fusion processes forge heavier atoms, such as neon, sulphur, silicon and iron. But eventually no new nuclear fusion reactions are possible and the final collapse begins. At this point so much energy is released that the star blasts itself apart in a supernova explosion which may be as bright as an entire galaxy of 100 billion stars. In this inferno even heavier atoms are created and dispersed into interstellar space to form gas clouds that eventually coalesce into the next generation of stars. Our bodies and all else around us are formed from atoms forged in the alchemical furnaces of the cosmos.

Diamonds in the Sky!

Galileo first pointed his telescope at the night sky just five lifetimes ago. His enhanced view of the heavens would revolutionize our understanding of the universe.

Starry Messenger

Wherever Galileo looked he made sensational discoveries. He could see mountains and craters on the Moon. Venus appeared as a crescent and showed phases like the Moon. There were dark spots on the face of the Sun and the Milky Way was composed of innumerable stars invisible to the unaided eye. Most spectacular of all were four moons that he watched dancing around Jupiter from night to night. Galileo raced to tell the world in a booklet called *Starry Messenger*, which was rushed into print early in 1610. This new vision of the cosmos transformed astronomy and shattered traditional ideas of the universe and our place within it (Figure 72). In the centuries since Galileo each technological advance in observing the heavens has produced further dramatic discoveries.

Alvan Clark & Sons

The first telescope factory in the United States was established in the 1840s by Alvan Clark, a portrait painter and engraver from a family of Cape Cod whalers. He was joined in the business by his

The Cosmic Mystery Tour. Nicholas Mee, Oxford University Press (2019). © Nicholas Mee.
DOI: 10.1093/oso/9780198831860.001.0001

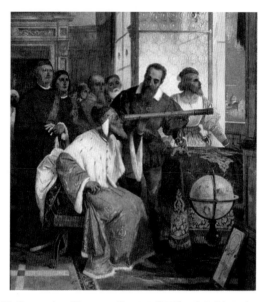

Figure 72 Fresco by Giuseppe Bertini (1858) of Galileo showing the Doge of Venice how to use the telescope.

sons Alvan Graham Clark and George Bassett Clark. Over the next half century, the company would grind lenses for some of the largest refracting telescopes ever made, five times setting new world records for their size. The biggest undertaking of all was a 40-inch lens for the Yerkes Observatory in Williams Bay, Wisconsin, which remains the largest refracting telescope in the world (Figure 73).

In 1861, Alvan Clark & Sons was commissioned to make an 18.5-inch telescope for the University of Mississippi. Due to the outbreak of the Civil War the telescope never reached its intended destination and in 1866 it was installed at the Dearborn Observatory of the University of Chicago. While testing the instrument on

Figure 73 Alvan Graham Clark with his assistant Carl Lundin polishing the lens for the Yerkes Observatory in 1896.

31 January 1862, Alvan Graham Clark looked at Sirius, the brightest star in the night sky. Twenty years earlier Friedrich Bessel had noticed a slight wobble in the position of Sirius and concluded this must be due to the gravitational attraction of an unseen companion. Clark became the first person to glimpse the companion star and confirm Bessel's prediction. Sirius is in the constellation Canis Major, one of the dogs accompanying Orion the Hunter. Since ancient times Sirius has been called the Dog Star, so its companion, Sirius B, is sometimes known affectionately as the *Pup*.

There is something rather odd about the Pup. The masses of the two components of Sirius can be determined from their orbit. Sirius A is twice the mass of the Sun, while Sirius B is almost the same mass as the Sun, so the brilliant Dog Star has just twice the mass of the faint Pup, yet it is incomparably brighter. The Hubble Space Telescope image in Figure 74 shows just how great the difference in luminosity is. The main star is

Figure 74 Sirius A and its white dwarf companion Sirius B or the Pup. (The spikes and rings are artefacts of the optics.)

Sirius A and its companion—the Pup—is the tiny dot close to the bottom left spike.

In 1915, the Pup was recognized as a white dwarf. These stars are remarkable because they are extremely hot—the temperature of the Pup is around 25,000 K—but very faint compared to other stars of the same mass. At a distance of 8.6 light years, the Pup is the closest white dwarf to us.

The Dog Star

Although the Pup is very difficult to spot, even with a good telescope, Sirius is a magnificent sight on a winter's night in the northern hemisphere. Sirius can be found by following the line of Orion's belt, as shown in Figure 75.

It takes 50 years for Sirius A and B to complete one orbit. Their average separation is comparable to the distance between the Sun and Uranus. At the moment they are approaching their widest separation, which will occur in 2019, so if you do fancy tracking

Figure 75 A guide to the star Sirius.

down the Pup with a telescope the next few years will offer the best opportunity.

Logic is the Beginning, Not the End of Wisdom!

Sirius B is notoriously difficult to see, because it is drowned out by the brilliance of Sirius A. The easiest white dwarf to observe is a member of a triple-star system known to astronomers as omicron 2 Eridani. It is also known as Keid, which derives from the Arabic for broken eggshell. These three stars are close stellar neighbours at about 16 light years distance. The main star Keid A is visible to the naked eye. It is orbited by a binary consisting of a white dwarf and an even fainter red dwarf that require a telescope to be seen (Figure 76). (Red dwarfs are ordinary stars, but lower in mass than the Sun. They are like very feeble versions of the Sun, generating energy by converting hydrogen into helium.) According to Gene Roddenberry, the creator of *Star Trek*, Spock's home planet Vulcan orbits Keid A.

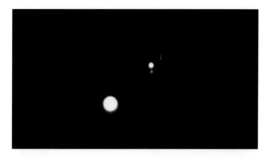

Figure 76 The large star is Keid A. It is orbited by a binary system composed of a white dwarf and an even fainter red dwarf. These stars are point-like when viewed from Earth, but they have been intentionally imaged slightly out of focus.

Credit: Image copyright 2008 Euan Mason.

The Puppy Dog's Tale

The Pup has the same mass as the Sun, yet it is extremely faint. Arthur Eddington argued in 1924 that this could only be explained by assuming that white dwarfs are much smaller than normal stars. He estimated the Pup to be similar in size to the Earth, which implies it is incredibly compact with a density around a million times that of water and 50,000 times that of gold.

A Golden Ring

We now know how white dwarf stars form. They are the dying embers of stars that have come to the end of their nuclear fusion generation. When the fuel runs out, the core contracts and the temperature rises. The outer layers of the star's envelope disperse into space to form a planetary nebula leaving the naked shrunken core radiating into space at a temperature of around 100,000 K when newly created. The Hubble Space Telescope image in Figure 77 shows a white dwarf at the centre of a famous planetary nebula

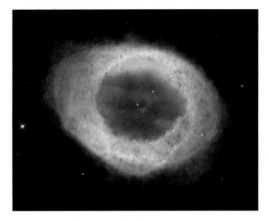

Figure 77 The Ring Nebula.

known as the Ring Nebula. The white dwarf is the dot close to the centre of the ring.

In about 10,000 years the planetary nebula will have dispersed into the background interstellar gas leaving the white dwarf shining faintly without its halo.

The Diamond in the Ring

The composition of a white dwarf depends on whether its nuclear fusion processes ceased after the hydrogen burning stage or after the helium burning stage. The first type of white dwarf is composed mainly of helium; the second type, which includes Sirius B, is formed of a mixture of carbon and oxygen. The white dwarf gradually radiates its heat into space and as it cools it crystallizes from the centre outwards. Figure 78 shows an artist's conception of the inside of a carbon–oxygen white dwarf such as Sirius B. These stars are the biggest diamonds in the cosmos.

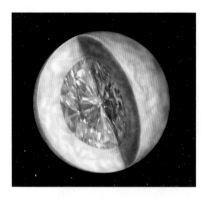

Figure 78 Carbon–oxygen white dwarf.

The greater the mass of a white dwarf, the more it is crushed by its own gravity, so unlike ordinary stars whose size increases with mass, the size of a white dwarf decreases with mass. A recent accurate measurement of Sirius B estimates its radius as 5,800 kilometres. By comparison the radius of the Earth is 6,371 kilometres, so, just as Eddington deduced over 90 years ago, Sirius B is the

Figure 79 A guide to the star Keid.

same mass as the Sun, but packed into a volume smaller than the Earth.

An Alternative *Star Trek*

If tracking down the Pup proves too tricky, you could try your luck with Keid, which is on the other side of Orion, as shown in Figure 79. You will still need a telescope to spot its white dwarf companion, but it could be the biggest diamond you will ever see.

From the Leviathan to the Behemoth

Alvan Clarke & Sons ground lenses for the biggest refracting telescopes in the world. But the real giants of the telescope kingdom are reflectors. Whereas refractors use lenses to focus light, reflectors use concave mirrors.

The Leviathan

William Parsons, the 3rd Earl of Rosse, built a monster telescope with a six-foot or 1.8-metre mirror weighing almost three tons at his home in Birr Castle in County Offaly, Ireland. Construction was completed in 1845. The Leviathan of Parsonstown, as it is known, remained in use until the end of the nineteenth century (Figure 80).

Figure 80 The Rosse Telescope *c.*1880.

The Cosmic Mystery Tour. Nicholas Mee, Oxford University Press (2019). © Nicholas Mee.
DOI: 10.1093/oso/9780198831860.001.0001

When the 4th Earl died in 1908 the telescope fell into disrepair and was partially dismantled, but it was not until 1917 that a bigger telescope was built anywhere in the world. In the late 1990s the Leviathan was renovated and a new mirror was installed.

Lord Rosse saw the heavens as they had never been seen before. He sketched these amazing sights and his drawings were widely circulated.

Camille Flammarion was a French astronomer and prolific author who wrote numerous books in the nineteenth century that brought science and astronomy to a wide readership. Flammarion's book *Astronomie populaire* published in 1879 was very successful and influential and a huge best-seller, over 100,000 copies were printed. Lord Rosse's sketch of the Whirlpool Galaxy was included in the book. In Lord Rosse's day it was thought to be a gaseous nebula within our own galaxy. We now know it is a beautiful spiral galaxy that is being perturbed by the gravitational influence of a smaller galaxy in its vicinity (visible to the right of the main spiral in Figure 81). It is certainly a stunning sight through a telescope.

Vincent van Gogh painted *The Starry Night* in June 1889, depicting the scene from his window in the Saint-Paul asylum in St. Rémy, Provence, where he admitted himself, seeking refuge

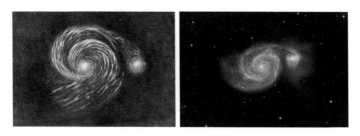

Figure 81 Left: Lord Rosse's sketch of the Whirlpool Galaxy. Right: Photograph of the Whirlpool Galaxy by Nik Szymanek.

Figure 82   *The Starry Night* by Vincent van Gogh (1889).

from his inner turmoil (Figure 82). But the striking feature of this wonderful painting is the sky. The cosmologist John Barrow makes a compelling case that the swirling patterns were inspired by Lord Rosse's sketch of the Whirlpool Galaxy, which Van Gogh may have seen in Flammarion's book or in contemporary newspapers following the great success of the book.

Fast Forward to Today

We are living through a golden age of astronomy. The advances in equipment in recent decades have been incredible—telescopes have come a long way since the days of Lord Rosse. The European Southern Observatory (ESO), funded by fourteen European nations, has constructed the most advanced optical observatory in the world, shown in Figure 83. The Very Large Telescope complex (VLT) sits at an altitude of 2,400 metres on a mountain top at Paranal in northern Chile. This site in the Atacama Desert is located in one of the driest regions on Earth. The incredibly low humidity and high altitude dramatically reduce the blurring

Figure 83 The Very Large Telescope complex in the Chilean Andes.

effects of turbulence in the atmosphere, giving the observatory a view of the night sky with unparalleled clarity. The VLT consists of four main telescopes, each with an 8.2-metre diameter mirror, along with four 1.8-metre auxiliary telescopes. The mirror of each auxiliary is the same size as the mirror of Lord Rosse's Leviathan.

In 1998, a competition was held to find appropriate names for the telescopes. The winner was a 17-year-old school girl, Jorssy Albanez Castilla, from Chuquicamata near the city of Calama, close to Paranal, who proposed in her winning essay that the telescopes should be named after celestial objects in the local Mapuche language. Following her suggestion, the four main telescopes are Antu, Kueyen, Yepun and Melipal. In Mapuche, Antu means Sun, Kueyen means Moon, Yepun means Venus and Melipal means Southern Cross.

Adaptive Optics

Even in the ideal setting of an Andean mountain top there is some atmospheric turbulence. To compensate for the distortion this produces in the starlight gathered by the huge telescope mirrors, the VLT uses adaptive optics. A laser is fired into the night sky and reflects off the upper atmosphere to create an artificial point of light or *guide star* that is continuously monitored through the telescope where a computer-controlled feedback mechanism deforms the telescope mirrors to keep its image point-like. This mechanism compensates for fluctuations in the atmosphere and enables the telescopes to produce much sharper images than

Figure 84 Star-forming region in the Large Magellanic Cloud.

would otherwise be possible. The images match those of a space-based telescope, but at a fraction of the cost. Furthermore, the four main telescopes and the four auxiliary telescopes can be operated as an interferometer in which the light from the telescopes is channelled through underground tunnels and combined to form an ultra-sharp image of incredibly high resolution.

The results are spectacular, as can be seen from the image in Figure 84, which shows a star-forming region in the dwarf galaxy companion of the Milky Way known as the Large Magellanic Cloud. The nebulae in the image have been produced by intense stellar winds from extremely hot newborn stars.

The Behemoth

The European Southern Observatory has started construction of what will be far and away the world's largest telescope. This new

Figure 85 An artist's impression of the ELT when complete.

behemoth is the Extremely Large Telescope (ELT). It is situated atop a mountain known as Cerro Armazones, 20 kilometres from the VLT complex, at an altitude of just over 3000 metres. The composite mirror of the telescope will be constructed from 798 hexagonal segments, each 1.4 metres wide, making the full mirror a gargantuan 39 metres in diameter (Figure 85). This will enable the telescope to collect about 15 times as much light as any existing telescope and 470 times that of the Leviathan of Parsonstown. The telescope will incorporate an advanced adaptive optics system using several lasers and actuators that can distort the shape of the mirrors a thousand times each second.

Construction of the Extremely Large Telescope began in early 2017 and the first observations are scheduled for 2024. Every advance in optical equipment has brought sensational discoveries about our place in the universe and the same will surely be true of the ELT. It will give astronomers the ability to search for the first generation of stars in the very early universe, it will enhance the search for Earth-like planets around distant stars and it will enable astronomers to probe the supermassive black hole at the centre of our galaxy, but, like its predecessors, it will also encounter the unknown.

The Crab and the Jellyfish

Fritz Zwicky was a Swiss astronomer who worked for most of his career at Caltech (California Institute of Technology) and the Wilson and Palomar Observatories (Figure 86). Zwicky was a creative and original thinker who ignored fashionable trends and pursued his own ideas. He often referred to himself as a lone wolf. He certainly had a rather misanthropic streak that left him with little sympathy for less talented colleagues, whom he described in the Introduction to his *Catalogue of Selected Compact Galaxies and Post-Eruptive Galaxies* as

> sycophants and plain thieves who have no love for any of the lone wolves who are not fawners and apple polishers, [they] doctor their observational data to hide their shortcomings and to make the majority of the astronomers accept and believe in some of their most prejudicial and erroneous presentations and interpretations of facts [and this is why they publish] useless trash in the bulging astronomical journals.

His ultimate contempt was reserved for those he referred to as 'spherical bastards', because they were 'bastards whichever way you looked at them'. It is perhaps no surprise that Zwicky had few academic friends. Justified or not, his tirades did not help his valuable insights to gain acceptance and his work was slow to receive the credit it deserved. Some of his imaginative ideas were mistaken, but more often he was on the right track decades ahead

The Cosmic Mystery Tour. Nicholas Mee, Oxford University Press (2019). © Nicholas Mee.
DOI: 10.1093/oso/9780198831860.001.0001

Figure 86 Fritz Zwicky

of the pack. He discussed gravitational lensing, dark matter and neutron stars in the 1930s. One of his most important breakthroughs relates to exploding stars.

Guest Stars

A star that appears, as if from nowhere, is known as a *nova*, meaning new star. Gradually the nova fades and disappears again. Such outbursts have been observed by astronomers for thousands of years. We now know they are produced by gigantic nuclear explosions on white dwarf stars. Some white dwarfs are like ticking time bombs and regularly produce brilliant outbursts. One such, RS Ophiuchi, flared up in 1898, 1933, 1958, 1967, 1985 and 2006.

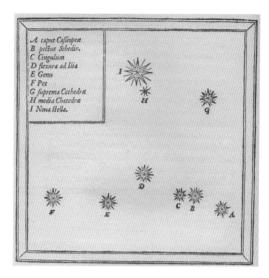

Figure 87 Tycho's chart showing the new star of 1572 labelled I. Stars FEDBG form the familiar 'W' of Cassiopeia.

The historical records also include very rare accounts of the sudden appearance of extremely bright stars, such as a new star witnessed by Tycho Brahe in 1572 (Figure 87) and one studied by Johannes Kepler in 1604. Several of these are recorded in the Chinese annals where they are referred to as *guest stars*. The brightest such guest star appeared on 1 May 1006 in the southerly constellation of Lupus (The Wolf). It was recorded all around the world—accounts survive from Japan, China, Iraq and even the monastery of St Gall in Switzerland. For several months it was so bright it was visible during the daytime, and the Egyptian astronomer Ali ibn Ridwan recorded that it was comparable in brightness to a quarter Moon. Guest stars such as these are much brighter than ordinary novae. But by 1930 none had been seen for well over 300 years.

The Vastness of the Universe

The true scale of the universe only became apparent with Hubble's work in the 1920s. Galaxies previously thought of as relatively nearby gas clouds were found to be vast collections of stars millions of light years distant. Our Milky Way was relegated to just one of hundreds of billions of galaxies in the universe (Figure 88).

Occasionally novae had been seen in other galaxies, such as one that occurred in the Andromeda Galaxy in 1885 (then known as the Andromeda Nebula). Zwicky and Walter Baade realized in 1931 that if these novae were actually in galaxies millions of light years distant, they must belong to the rare class of superbright stellar outbursts like Tycho's star. Zwicky and Baade labelled these spectacular explosions *supernovae*, arguing they were so much brighter than novae, they must have a different origin and arise from even more extreme cosmic events (Figure 89). Soon the two pioneer astronomers found examples in other distant galaxies and instigated systematic searches for more.

Figure 88 Detail of the Hubble Ultra Deep Field. Every object in the image is a separate galaxy.

Figure 89 Supernova SN1994D on the outskirts of galaxy NGC4526 about 50 million light years away. The supernova (to the bottom left of the photograph) is as bright as the core of the galaxy, which could contain 100 billion stars.

Cosmic Fireworks

Zwicky and Baade began their studies of supernovae just before the birth of nuclear physics when our understanding of stars was in its infancy. We now know that stars generate energy through nuclear fusion reactions. For most of a star's life these reactions steadily convert hydrogen into helium in the star's core and the energy this releases supports the star against its tendency to collapse under gravity.

Very massive stars burn their nuclear fuel at a phenomenal rate and when the hydrogen in their core runs out, the core contracts and the temperature rises until a new round of nuclear reactions is triggered. Helium is converted into carbon and oxygen, and in the most massive stars this is followed by fusion processes that create the nuclei of heavier atoms such as neon, magnesium, silicon, sulphur and iron. But eventually no further nuclear reactions are possible and in stars of several solar masses the final collapse begins; this liberates huge amounts of energy, and the star blasts itself apart in a supernova explosion.

Figure 90 The Jellyfish Nebula, known more formally as IC 443, is the remnant of a supernova that would have been seen several thousand years ago.
Credit: © Alessandro Falesiedi.

When the smoke clears the core of the star may have been transformed into an object with the density of an atomic nucleus. These weird objects are known as neutron stars. In stars of even greater mass the core collapse doesn't stop. The result is a black hole.

The most famous supernova remnant is the Crab Nebula (see Figure 94 in the chapter 'The Ultimate Heavy Metal Space Rock'); it is the highly energetic remains of a supernova seen in 1054. The stunning image in Figure 90 shows a much older supernova remnant known as the Jellyfish Nebula. The Crab and Jellyfish Nebulae are both known to contain a rapidly spinning neutron star, which is the super-dense collapsed core of the star that exploded.

Figure 91 Remnant of the supernova of 1006. This false colour image combines data from a wide range of the electromagnetic spectrum, X-ray data in blue, optical data in yellow and radio data in red.

There is another important class of supernovae produced by a different type of event—a runaway thermonuclear explosion of a white dwarf triggered by the detonation of large amounts of material accumulated from the outer layers of a companion star. These enormous blasts are thought to completely obliterate the white dwarf leaving nothing but a rapidly expanding cloud of material stuffed with heavy elements cooked up in the explosion. The brilliant supernova recorded around the world in 1006 was created in this way. Its remains were tracked down in 1965 and are shown in Figure 91. All that is left is an extremely hot gas cloud expanding at 10 million kilometres an hour. It is now about 60 light years across and lies at a distance of 7000 light years from us.

The Ultimate Heavy Metal
Space Rock

It is fifty years since the Summer of Love. During that summer a young Cambridge graduate student working on a newly developed radio telescope designed by Martin Ryle and Anthony Hewish noticed a bit of 'scruff' in her readout. Jocelyn Bell looked closer and realized there was something strange going on (Figure 92).

Figure 92 Jocelyn Bell.

The Cosmic Mystery Tour. Nicholas Mee, Oxford University Press (2019). © Nicholas Mee.
DOI: 10.1093/oso/9780198831860.001.0001

The signal was repeating every 1.337 seconds precisely and it could not be from any local source as it was moving across the sky with the Earth's rotation. There was no obvious astronomical phenomenon that could generate such regular and rapid pulses (Figure 93). Her first thought was that this might be a signal from an alien civilization. Provisionally the radio source was given the half-serious acronym LGM1 (Little Green Men 1). (It has since become known as CP1919 or PSR1919+21.)

Bell later recalled: 'we did not really believe we had picked up signals from another civilization, but obviously the idea had crossed our minds and we had no proof that it was an entirely

Figure 93 The periodic signal of the first pulsar discovered by Jocelyn Bell in 1967.

natural radio emission. It is an interesting problem—if one thinks one may have detected life elsewhere in the universe, how does one announce the results responsibly?'

The aliens were soon discounted as Bell discovered a second radio source pulsing away in the night sky. One alien civilization would be an incredible discovery, but two alien civilizations communicating in this way was simply too improbable. There had to be a natural explanation. In the meantime these objects were dubbed *pulsars*. When Hewish and Bell announced their discovery in 1968, Thomas Gold suggested they might be rapidly spinning *neutron stars*. Fritz Zwicky and Walter Baade had proposed the existence of such objects in 1934, just two years after the discovery of the neutron. Their prophetic idea was that following a supernova blast the remains of the exploding star would be compressed into an object with the mass of a star but as dense as an atomic nucleus and composed entirely out of neutrons. Gold argued that a neutron star would be spinning rapidly and its intense magnetic field would generate beams of radiation much like a rotating beacon. Initially, his theory was dismissed by the scientific community and he was even refused permission to present it to the first international conference on pulsars. Everything changed later in the year, however, when a pulsar with a period of 33 milliseconds was discovered in the Crab Nebula, the remnant of a supernova seen in the year 1054 (Figure 94).

Ryle and Hewish were awarded the 1974 Nobel Prize in Physics for the discovery of pulsars. Jocelyn Bell did not receive the prize despite her leading role in the discovery.

Cosmic Lighthouses

Neutron stars are incredibly compact stellar remnants with a mass greater than that of the Sun compressed into a ball with

Figure 94 The Crab Nebula.

a diameter of around 20–30 kilometres. They are as dense as an atomic nucleus and spin at an unbelievable rate, typically completing a full revolution in a fraction of a second.

The diffuse magenta-coloured area in Figure 95 is a false colour X-ray image of a supernova remnant. The green comet-like object towards the bottom right contains a pulsar that is moving at high speed and appears to be racing away from the site of the supernova. It is believed that the supernova underwent a lopsided explosion that sent the resulting neutron star careering on its way. (The two bright stars in the image are in the same line of sight, but are not thought to be related to the supernova event.)

Astronomers have now studied neutron stars for half a century. Much is known about them, but many mysteries remain. They are indeed created in supernova explosions and are believed to spin at close to one thousand times a second when newly created. As Gold first suggested, this generates a huge magnetic

Figure 95 Supernova remnant. The image is a composite of X-ray, infrared and optical data.

field a trillion times that of the Earth. This in turn generates electric fields that accelerate electrons and other charged particles, producing two intense beams of radiation that blast outwards from the magnetic poles of the neutron star (Figure 96). Just as on Earth, the magnetic poles are not perfectly aligned with the rotation axis, so the beams sweep around the heavens like a cosmic lighthouse. Radio astronomers, such as Jocelyn Bell, detect a pulse of radio waves once every rotation when the pulsar beam points in our direction.

The spinning magnetic field acts as a brake on the rotation of the neutron star, and the pulsar that it generates transmits the lost rotational energy to the surrounding nebula. Gradually the rate of rotation decreases. The neutron star within the Crab Nebula formed almost one thousand years ago and now spins about thirty times a second. Calculations show the amount of rotational energy it is losing matches the energy required to illuminate the nebula. So pulsars have a limited lifespan. In around one million years the neutron star's rotation period will have increased to about one

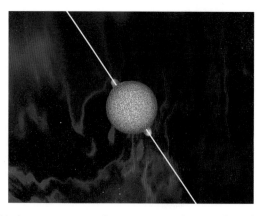

Figure 96 A representation of a neutron star showing the pulsar beams emanating from both magnetic poles.

second, by which time there will be insufficient energy to power the pulsar and it will disappear from view.

A neutron star in a binary system may accrete material from its companion, which is quite likely to happen if the companion sheds its outer layers as it ages or inflates into a red giant. As material spirals onto its surface, the neutron star is spun up and the long-dead pulsar may come back to life. This is believed to be the origin of *millisecond pulsars*, which are observed to have extremely short periods. Currently, the fasting spinning pulsar we know of is PSR J1748-2446ad, which rotates an incredible 716 times a second.

Exotic Materials

The composition of neutron stars is more complicated than their name might suggest. Their radius is just 10–15 kilometres and their structure changes as we move inwards, as indicated in Figure 97. They are thought to have a hot plasma atmosphere just a few centimetres thick surrounding an outer crust of

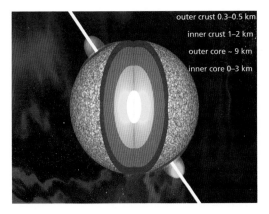

outer crust 0.3–0.5 km

inner crust 1–2 km

outer core ~ 9 km

inner core 0–3 km

Figure 97 The internal structure of a neutron star.

white-dwarf-like matter consisting of heavy nuclei bathed in a sea of electrons, with a density around a million times that of water.

Moving inwards the density increases rapidly. When we reach the inner crust the proportion of neutrons in the nuclei dramatically increases along with the density of free neutrons until we reach a critical threshold at around 100 trillion times the density of water.

We now enter the outer core, which consists almost exclusively of neutrons plus a small quantity of protons, electrons and muons. No one knows the physical structure of the material forming the inner core at the heart of a neutron star. Various exotic possibilities have been proposed. These include mind-boggling suggestions such as densely packed particles known as strange baryons, which are like heavy versions of protons and neutrons but containing strange quarks; a Bose–Einstein condensate of particles known as pions and kaons that are composed of quarks and antiquarks; or possibly some sort of quark–gluon plasma, which is an extreme state of matter currently being explored in the Large Hadron Collider.

There is an upper limit to the mass of a neutron star that is thought to be somewhere between two and three times the mass of the Sun. Nothing can stop the total collapse of a neutron star whose mass exceeds this limit—it is doomed to undergo an inexorable implosion and form a black hole.

A Cultural Icon

In 2016 the Institute of Physics decided to rename its Very Early Career Female Physicist Award the Jocelyn Bell Burnell Medal and Prize in recognition of the achievements of Professor Dame Jocelyn Bell Burnell and her discovery of pulsars during her PhD research.

The signal discovered by Jocelyn Bell has become a cultural icon thanks to the designer Peter Saville, who in 1979 used it as the cover art for the Joy Division album *Unknown Pleasures* (Figure 98).

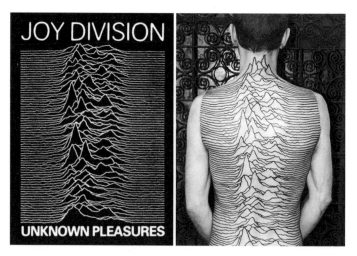

Figure 98 Left: Album cover for Joy Division's *Unknown Pleasures*. Right: Body art based on the Joy Division cover artwork.

The beat of the first known cosmic lighthouse has been transformed into one of the most recognizable images in rock music. It reminds me of the track followed by the needle through the groove of a vinyl LP. Incidentally, Ian Curtis, lead singer and lyricist of Joy Division, grew up in the Cheshire town of Macclesfield close to the Jodrell Bank radio observatory that is now a world-leading centre for research into pulsars. At Jodrell Bank they have even turned the pulses of these spinning space rocks into audio files.

Pan Galactic Gargle Blaster

Astronomers Joseph Taylor and Russell Hulse searched systematically for pulsars with the giant Arecibo radio telescope in Puerto Rico in 1974 and found many new examples of these extraordinary objects. The radio pulses in pulsar signals are received with incredible regularity. But amidst the data was one whose behaviour seemed very strange. Curiously, the intervals between the pulses would increase for a while and then decrease again in a regular pattern that would repeat every seven and three-quarter hours.

Hulse and Taylor had discovered a binary system consisting of two orbiting neutron stars and the pulsar belongs to one of them. As the neutron star approaches Earth the interval between the arrival of its pulses steadily decreases, then as the neutron star recedes the interval increases again. This is the familiar Doppler effect we hear with moving sirens. When the siren moves towards us its pitch rises and when it moves away its pitch falls. If the other neutron star has a pulsar, it has not been detected, so it never points in our direction.

The binary neutron star was a wonderful find because the regular pulses have enabled astronomers to study the motion of this binary system with great precision and determine its vital statistics. We know that both neutron stars are around one and a half solar masses and their separation is very small by cosmic standards. The neutron stars approach each other to a distance slightly greater than the radius of the Sun and they recede to a widest separation almost five times this distance, so the orbit is a

The Cosmic Mystery Tour. Nicholas Mee, Oxford University Press (2019). © Nicholas Mee.
DOI: 10.1093/oso/9780198831860.001.0001

highly eccentric ellipse. Each orbit is completed in just 7.75 hours.
It is remarkable to have such detailed information about the
paths followed by objects that are a vast 20,000 light years away—
about a billion times further away than the Sun.

But the Hulse–Taylor neutron star system has even more to
offer. The Solar System is very sedate, with the planets serenely
orbiting the Sun for aeon upon aeon. Kepler's first law states that
each planet follows an ellipse whose axis points in a fixed direction
in space and the planet endlessly follows the same ellipse. Kepler
deduced the law from Tycho Brahe's observations and it was published in 1609, decades before Newton provided an explanation.

Newton's theory of gravity accounts for the motion of each
planet with great accuracy. Only in the case of Mercury was
there a small hiccough that required a better theory. The slight
discrepancy in Mercury's orbit was spotted in the middle of the
nineteenth century and it remained a mystery until Einstein
provided the solution in 1915 in one of his greatest triumphs.
General relativity predicts that the planetary orbits don't quite
close up; they precess. The direction of the axis of the ellipse
gradually rotates to form a rosette shape, as shown in Figure 99.

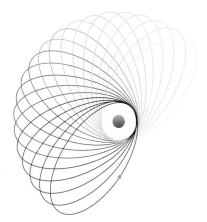

Figure 99 The precession
of Mercury's orbit (greatly
exaggerated).

But the rate of precession is minuscule unless the planet is moving rapidly in a very strong gravitational field. Mercury is the closest planet to the Sun. It is deeper in the Sun's gravitational well and it races around much faster than the other planets, so the relativistic corrections to its motion are much greater. But the effect is still very small.

The Laboratory at the End of the Universe

The Hulse–Taylor neutron star system is a far more extreme gravitational environment than the Solar System, so the relativistic effects are much greater and the pulsar offers a built-in precision timekeeper providing astronomers with a great physics laboratory in which they can put Einstein's theory to the test. When the paths of the neutron stars were mapped out in detail, as expected, their orbit didn't quite match the Newtonian prediction, but it agreed with Einstein's theory exactly. This was a wonderful confirmation of our modern understanding of gravity and it is still one of the most significant successes of general relativity. Even so, this effect had already been seen with Mercury many years earlier, so it wasn't quite headline news.

Shrinking Orbits

But there was something else that was even more exciting. The orbital period of the Hulse–Taylor binary system can be measured with exquisite precision. Although it is currently about seven and three-quarter hours, it is decreasing by 76.5 microseconds every year.

General relativity provides the explanation. As the two neutron stars whirl around each other they generate gravitational waves—ripples in the fabric of space. The energy that is radiated away in these waves causes the orbit of the binary system to shrink. When physicists calculated the expected rate of orbital decay due to gravitational wave emission the prediction of general relativity matched the observations to perfection. So decades before researchers detected gravitational waves directly at LIGO their existence was confirmed by this extraordinary star system. Hulse and Taylor were awarded the Nobel Prize in Physics in 1993 for their discovery and subsequent analysis of the binary neutron star.

The orbit of the neutron stars shrinks by about 3 metres every year due to the emission of gravitational waves. In around 300 million years the two neutron stars are scheduled for a close encounter. The outcome is sure to be incredibly violent.

A Blast from the Past

The Nuclear Test Ban Treaty was signed in 1963. To monitor compliance with the treaty, in 1967 the United States launched a series of satellites carrying instruments that would detect gamma rays—the telltale signatures of nuclear explosions. Occasionally the satellites did pick up a worrying flash of gamma rays and these gamma ray bursts, as they are known, were immediately put under investigation. It soon became clear, however, that they originate in deep space and not some illicit weapons programme. By 1973 the research was declassified and astronomers were given the chance to investigate their origin. As the gamma ray bursts typically last just a few seconds uncovering their secrets proved quite a challenge. Success has depended on the rapid deployment of telescopes to study the afterglow following the detection of a burst by a gamma ray detector aboard a satellite. We now know they are produced by some of the most violent cataclysms in the universe. They are rare, but so powerful they can be detected from immense distances.

Gamma ray bursts have been divided into two categories. Most of the signals last for a few seconds and are known as long gamma ray bursts. The other category—the short gamma ray bursts—last for less than two seconds.

Staring down the Barrel of a Ray-gun

Astronomers now think they know the source of the short gamma ray bursts. On 3 June 2013 the gamma ray telescope aboard NASA's Swift satellite (Figure 100) picked up a gamma ray burst lasting just a tenth of a second. Nine days later the Hubble Space Telescope took up the search for the source of the gamma radiation. Hubble found a faint glow in a galaxy 4 billion light years away where an event that occurred when Earth was in its infancy had produced the gamma ray burst.

Figure 100 NASA's Swift satellite.

Last Tango in Deep Space

The Hubble images indicate the short gamma ray burst was generated by a type of stellar explosion known as a *kilonova*. What sort of drama produces a kilonova, you might ask? They are thought to be generated by the merger of two neutron stars that have reached the climax of their cosmic dance. This will be the ultimate fate of the neutron star system discovered by Hulse and Taylor. We now know much more about these rare but rather important cataclysms, as we will see in the next chapter.

Cosmic Spacequakes

Long ago in the year AD 132 the imperial astronomer Zhang Heng designed an earthquake detector. The *History of the Later Han Dynasty* reports that his ingenious invention would alert the Chinese emperor to catastrophic seismic events in remote outposts of the empire. Zhang Heng's seismoscope is described as a bronze vessel two metres in diameter with eight dragon heads mounted around its circumference. A reconstruction is shown in Figure 101. Each dragon clasps a small metal ball in its teeth, while the open mouth of a bronze toad gapes wide below. A faint tremor from a distant earthquake causes a rod within the vessel to overbalance

Figure 101 A reproduction of Zhang Heng's seismoscope.

The Cosmic Mystery Tour. Nicholas Mee, Oxford University Press (2019). © Nicholas Mee.
DOI: 10.1093/oso/9780198831860.001.0001

pulling a lever that opens the mouth of the dragon facing towards the earthquake. Its ball is released and falls clanging into the waiting mouth of a toad.

From Earthquakes to Spacequakes

Fast forward almost two thousand years and gravitational wave observatories now routinely detect spacequakes in far-flung regions of the universe. In August 2017 the two LIGO detectors in the United States were joined by the newly upgraded VIRGO detector near Pisa in Italy. By determining the arrival time of faint cosmic rumbles at each detector with split second accuracy it is possible to get a good fix on the direction of the source of the gravitational ripples. This is valuable information as it enables astronomers to seek any visible sign of the catastrophic blast that produced the gravitational waves. On 17 August the three instruments captured an unmistakable signal, quite different to the four previous events detected by LIGO. It was catalogued as GW170817. Just 1.7 seconds later NASA's Fermi Gamma-Ray Space Telescope detected a short gamma ray burst emanating from the same region of sky and telescopes around the world were marshalled to locate the glowing embers of the event that produced the gamma radiation. Within twelve hours the source of the gamma ray burst was tracked down.

This was the first discovery of an optical counterpart for a gravitational wave signal and it has become one of the most observed events in the history of astronomy. The gravitational waves and short gamma ray burst were generated by a kilonova 138 million light years away in a galaxy known as NGC 4993 (Figure 102). Kilonovae form a new class of stellar explosions intermediate between novae and supernovae. They are around one thousand times brighter than a nova and this is the reason

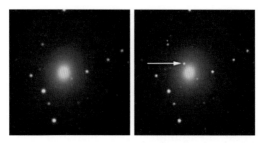

Figure 102 Left: Overexposed image of the galaxy NGC 4993 before the kilonova explosion. Right: The arrow indicates the kilonova.

for their name. Nonetheless, they are just one-thousandth to one-hundredth of the luminosity of a supernova. We now know a kilonova is the spectacular explosion produced when a pair of neutron stars collide and merge.

Neutron Star Collisions

Neutron stars are like gigantic atomic nuclei—the entire mass of a star is compressed to nuclear densities. They are among the strangest objects in the cosmos. Typically, one and a half solar masses is packed into a sphere just 20 or 30 kilometres in diameter. Tiny by cosmic standards, but incredibly dense, the mass of a tea-spoonful of neutron star far exceeds a billion tonnes. Try stirring that into your cup of tea!

The neutron stars whose collision generated the GW170817 event are presumed to have had a similar history to the Hulse–Taylor binary system explored in the previous chapter. Over the course of millions of years they would have gradually spiralled together as they lost energy due to the emission of gravitational waves. The amplitude of these waves was too small to detect until

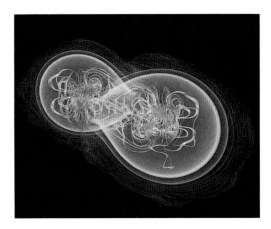

Figure 103 Artist's impression of merging neutron stars.

they came almost within touching distance 100 seconds before their fatal encounter (Figure 103).

Neutron stars are composed of an exotic form of matter consisting largely of neutrons. Such material only exists within the extreme gravitational stranglehold of a neutron star. Any fragment smashed free in a collision would be extremely unstable and undergo immediate radioactive decay with dramatic and violent consequences. Neutrons would transform into protons with the emission of electrons and neutrinos. Neutron-rich heavy nuclei would form and rapidly decay into more stable lighter nuclei in a blaze of gamma radiation. This is the origin of the short gamma ray burst detected by the Fermi satellite. The ongoing radioactive decay of the material emitted in the merger produces the visible afterglow of the kilonova (Figure 104). The neutron stars that merged in the GW170817 event are believed to have undergone the ultimate collapse and formed a black hole.

Figure 104 Artist's conception of a kilonova.

The Birth of Heavy Metal

Neutron star collisions might answer one of the mysteries of the cosmos. How were heavyweight atoms such as platinum and gold created?

Massive stars generate energy through a sequence of nuclear fusion processes that ultimately transform the core of the star into iron and nickel. Eventually, no further energy can be generated by fusion reactions and the core collapses, triggering the detonation of the star as a supernova. The exploding core of iron and nickel is bathed in a vast flux of high energy protons and neutrons and this was long thought to explain the origin of heavier nuclei all the way up to uranium and plutonium. However, recent computer simulations suggest the extreme conditions necessary for nucleosynthesis may not last long enough for the creation of elements beyond silver and its neighbours in the Periodic Table. This presents us with a conundrum: Where do all the really heavy elements come from?

David Eichler, Mario Livio, Tsvi Piran and David Schramms suggested in 1989 that neutron star collisions might provide an alternative process in which heavy elements are synthesized. This attracted little support at the time as it was assumed that such

Figure 105 Tutankhamun's magnificent gold mask.

events would be too rare to account for the quantities of gold, uranium and other elements we find in the galaxy (Figure 105).

Interstellar Gold Rush

Computer models of neutron star mergers, such as the one that produced the GW170817 signal, show that around 20,000 times the mass of the Earth could be ejected in these events. This material would be blasted out at about one-fifth of the speed of light and dispersed far and wide throughout the galaxy. It would be in the form of heavy elements including about ten parts per million of gold nuclei, so the total amount of gold dust created in a kilonova would be about one-fifth of the mass of the Earth. Observations of the recent kilonova suggest that the computer models are correct.

But are these events frequent enough to account for the observed amounts of heavy elements?

Tightening the Net

The big question is whether heavy elements are produced in relatively small quantities in unusual events—supernovae—or whether they are produced in extremely large quantities in very rare events—neutron star collisions.

Support for the latter possibility comes from a recently discovered dwarf galaxy known as Reticulum II. Discovered in 2015 and located in the obscure southern constellation of Reticulum or the Little Net, this galaxy is a close companion of the Milky Way, just 97,000 light years away. Most dwarf galaxies contain little, if any, heavy elements. By contrast, significant amounts of the heavy elements are present in the stars and interstellar gas of Reticulum II.

Supernova explosions are unusual events. It is estimated that one occurs in the Milky Way Galaxy every 30 years or so, but most are hidden from us by interstellar dust clouds. Even in the feeblest of dwarf galaxies, we would expect a supernova at least once every 100,000 years, so if the heavy elements are produced in supernovae, over the 13.8-billion-year history of the universe the heavy elements would build up to significant levels. Neutron star collisions could be so rare, however, that most dwarf galaxies have never hosted such an event and this could explain why most dwarf galaxies do not contain heavy elements. It looks as though the Reticulum II dwarf galaxy is simply the winner of the neutron star collision lottery. All the heavy elements that it contains were probably produced in a single such event. This provides important circumstantial evidence that long ago most of the heavy elements in our environment were also produced in this way.

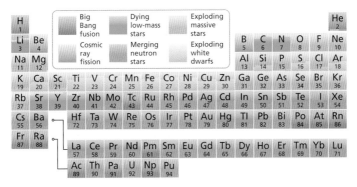

Figure 106 The origin of each element of the Periodic Table. Gold is element 79, symbol Au.

The chart in Figure 106 indicates the origin of each element in the Periodic Table according to modern astrophysics. (The chart only shows the final process in which a particular nucleus is forged. Most nuclei are created following a sequence of possibly very different processes. For instance, boron nuclei are created when cosmic rays hit carbon nuclei and dislodge a proton, but the carbon nucleus would have been created in a star or supernova explosion.)

Your Gold Ring

It is an amazing fact that all the atoms in your body, other than the hydrogen atoms, must have been created within a star. It is perhaps even more incredible that the atoms in the gold ring on your finger were created in a neutron star collision.

And what of the long gamma ray bursts? That is a topic for the next chapter.

Doctor Atomic and the Black Hole

Freedom

In 1970 NASA placed its first X-ray telescope in Earth orbit. The satellite was launched from Kenya on 12 December, the seventh anniversary of Kenyan independence from Britain. In honour of their hosts, NASA named the satellite Uhuru, Swahili for freedom. One of its first targets was a powerful X-ray source located in the constellation Cygnus the Swan and designated Cygnus X-1. It proved to be a very interesting celestial object—the first stellar mass black hole ever identified.

A black hole is a region of space containing matter that has been squeezed to an incredible density. It is thought that stars of more than about twenty solar masses must end their lives in a cataclysmic collapse that results in the creation of a black hole. The core of the star is crushed in the death throes of the star and all that remains is a warped region of space whose gravitational attraction is so strong that once inside nothing can escape, not even light. The dividing line between the black hole and the rest of the universe is known as the black hole's *event horizon*. This is a spherical boundary and once crossed there is no return.

On Continued Gravitational Attraction

The theoretical physicist John Wheeler first named these weird objects black holes in 1967, but their existence was suspected

The Cosmic Mystery Tour. Nicholas Mee, Oxford University Press (2019). © Nicholas Mee.
DOI: 10.1093/oso/9780198831860.001.0001

Figure 107 A scene from *Doctor Atomic*: Oppenheimer is on the left, the atom bomb is on the right.

almost three decades earlier when Robert Oppenheimer and his student Hartland Snyder examined the implications of general relativity and showed it demands the existence of objects whose gravity is so intense even light cannot escape. This startling conclusion formed part of their paper 'On Continued Gravitational Attraction' submitted to the journal *Physical Review* on 1 September 1939, the day that Germany invaded Poland, triggering the start of World War II. Three years later Oppenheimer was appointed head of the Manhattan Project's secret weapons laboratory in Los Alamos, New Mexico, tasked with developing the world's first nuclear weapons.

Oppenheimer is the eponymous hero of the opera *Doctor Atomic* by John Adams and Peter Sellars first performed in 2005 (Figure 107). The action revolves around preparations at Los Alamos in 1945 for the first nuclear bomb test codenamed Trinity.

Cygnus X-1

Astronomers have been studying Cygnus X-1 for over forty years with instruments sensitive to light in various wave bands—radio, optical and X-ray. The distance to Cygnus X-1 was recently determined using the US National Radio Observatory's Very Long Baseline Array as 6,070 light years and this figure has helped pin down the properties of the black hole. It is an impressive 14.8 solar masses and spins an incredible 800 times per second.

Astronomers believe Cygnus X-1 began as a binary system containing two very massive, but otherwise ordinary stars bound by their mutual gravitational attraction. The heavier of the two would have burnt its nuclear fuel much faster than its partner. About six million years ago the fuel ran out and with the shutdown of its nuclear power source the star underwent its terminal collapse and became a black hole. Although the original star was millions of kilometres in diameter, the black hole is tiny, just a few kilometres across, but with a mass almost fifteen times that of the Sun. The black hole and its companion star remain gravitationally bound and continue to perform their orbital pirouette.

Figure 108 An artist's conception of the Cygnus X-1 system.

The black hole is surrounded by an *accretion disc* of material that is gradually being drawn from its neighbour, as shown in Figure 108. As this material swirls inwards towards the black hole it is heated to ultra-high temperatures, causing it to emit the intense X-rays first detected by Uhuru.

Into the Abyss

Long gamma ray bursts are thought to signal the gravitational collapse of a huge star, such as the parent of Cygnus X-1. Black holes are often portrayed as gaping like the jaws of Hell. It is certainly true that once inside it is impossible to escape, but getting inside may not be that easy. Black holes are tiny by cosmic standards—the diameter of Cygnus X-1 is less than one millionth of the diameter of its parent star. Squeezing an entire bloated star into such a small volume proves rather difficult, so black holes are very messy eaters. Much of the stellar-sized meal narrowly avoids entering the abyss and shoots out at the poles. This plasma is compressed beyond nuclear densities and focused into two beams that blast outwards at almost the speed of light, generating intense gamma ray beams that race across the universe. A civilization on the other side of the cosmos that happens to be looking directly down the barrel of this gamma-ray gun may eventually pick up a brief trace of radiation signalling the death of a mighty star and the formation of a black hole.

Quasars

The 1960s saw the discovery of a new class of astronomical objects named *quasars* (quasi-stellar objects). Although quasars looked point-like and therefore similar to stars, the lines in their spectra showed far greater red shifts than any other known objects, indicating they were located at vast distances. To be visible at such distances their energy output had to be stupendous, much greater than the combined output of all the stars in an entire galaxy. We now know they are produced by extremely violent activity at the heart of distant galaxies and blaze away with a *trillion* times the luminosity of the Sun. More astonishing still, their brightness often fluctuates over periods as short as a few hours, which implies their active regions are minuscule by galactic standards—just a few light hours across—about the size of the Solar System.

The source of all this energy was a great mystery. The astrophysicist Donald Lyndon-Bell suggested in the late 1960s that only one explanation was possible. He proposed that the enormous energy released by a quasar must be due to material falling into a black hole situated at the centre of a galaxy, as no other object could conceivably produce such a vast output of energy from such a small region of space. To account for the incredible luminosity of a quasar, however, these black holes had to be millions or even billions of times the mass of the Sun and with an appetite to match, devouring several stars' worth of material every year. Cygnus X-1 is an impressive beast, but if Lyndon-Bell was correct, it would be a mere baby compared to the monsters inhabiting the central regions of galaxies. Is it really credible that such outlandish objects exist? Before passing judgement, it would be nice to have some evidence from closer to home. In the next chapter we will zoom in to the centre of our own galaxy and take a look.

Supermassive Black Holes

The constellation Sagittarius is home to the beautiful region of the night sky shown in Figure 109. Some of the stars have been joined by lines to highlight an asterism known to amateur astronomers as the *teapot* which forms part of the constellation. This is a rather apt name as the many nebulae and gas clouds located towards the centre of the galaxy appear as steam rising from the spout of the teapot. The precise centre of the galaxy is indicated by an 'X' in the image. It is about 25,000 light years away.

Figure 109 X marks the spot of the centre of the galaxy, a region known to astronomers as Sgr A*.

The Cosmic Mystery Tour. Nicholas Mee, Oxford University Press (2019). © Nicholas Mee.
DOI: 10.1093/oso/9780198831860.001.0001

Where the Action Is!

Radio astronomers have named this region Sgr A*, an abbreviation that means the most powerful source of radio signals in the constellation Sagittarius. The star * is added to emphasize the special nature of the object. It is where the action is in the Milky Way Galaxy. Our immediate cosmic neighbourhood is incredibly quiet—the Sun is surrounded by vast oceans of space, it is over four light years to the nearest star. By contrast, within one light year of the centre of the galaxy there are, perhaps, a million stars. These include lots of burnt-out stellar remnants such as neutron stars and black holes, as well as many luminous blue supergiants.

The Innermost Heart of the Galaxy

German astronomer Reinhard Genzel studied the innermost heart of the galaxy in the early 1990s using the European Southern Observatory's 3.5-metre New Technology Telescope in Chile. His observations showed the stars at the centre of the galaxy are moving extremely fast, and the closer to the centre the faster they are travelling. This suggests there is a very high concentration of mass right at the centre. Furthermore, the location of this point appears to be fixed while all else whirls frantically around it.

Genzel's observations were followed up by American astronomer Andrea Ghez and her team with the two 10-metre Keck telescopes in Hawaii. The stars right at the centre of the galaxy are moving so quickly that over the course of just a few years they have traced out significant segments of their orbits. The closest neighbours to Sgr A* are racing around at up to 5 million kilometres per hour. As well as tracking their motion across the sky, it is possible to measure their velocity towards or away from us

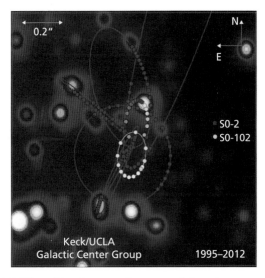

Figure 110 The orbits of stars at the centre of the galaxy as mapped by Andrea Ghez and her team.

through the Doppler shift of their light. This has enabled Ghez and her team to calculate accurate trajectories of these stars in three dimensions. One such star designated SO-2, which takes 15.5 years to complete its highly eccentric orbit, has been monitored over the course of an entire orbit. Another star known as SO-102 has an even smaller 11.5-year orbit, as shown in Figure 110. The enormous speed of these stars indicates they must be orbiting an absolutely immense mass.

The Mystery Object

Given the information gathered by Ghez about the orbits of the stars at the centre of the galaxy, the mass of this mysterious object is easily calculated. From the size of the stars' orbits and the time taken to complete each circuit, we know it is around 4 million solar masses. But observations show it would easily fit within the Earth's orbit around the Sun. There is only one possible conclusion—it is a supermassive black hole. The event horizon of a black hole of 4 million solar masses has a diameter of 24 million kilometres or twenty times the diameter of the Sun.

The existence of such a massive black hole at the centre of our galaxy gives credibility to the idea that monster black holes are responsible for the energy output of distant quasars and this is now the generally accepted explanation. Quasars were much more common in the early universe. Astrophysicists believe that all galaxies, including our own Milky Way, went through a quasar phase in their early history before settling into a quieter existence once their central black hole had consumed all the material in its vicinity. The supermassive black hole at the centre of our own galaxy is now relatively quiet and although it is big it is certainly no record breaker.

Bigger and Bigger

There is a much larger supermassive black hole in our galactic neighbourhood. The giant elliptical galaxy M87 in the Virgo cluster of galaxies, 53.5 million light years away, has a distinctive jet of material spewing from its core. In visible light, this jet extends around five thousand light years into intergalactic space. But radio telescopes can detect lobes of material emitted by the jet reaching as far as 250,000 light years from the galaxy.

Located at the heart of the M87 Galaxy is a supermassive black hole of 6.5 billion solar masses and this is believed to be the origin of the jet (Figure 111). (It is assumed that a second jet points in the opposite direction, but we can only see the one directed towards us.)

In all probability all galaxies contain a supermassive black hole at their centre and it seems the bigger the galaxy the bigger the black hole. Within the Coma cluster of galaxies 300 million light years

Figure 111 The diffuse amber sphere is the giant elliptical galaxy M87. The jet emanating from its centre is thought to originate at the supermassive black hole at the centre of the galaxy.

away there is a gargantuan supergiant elliptical galaxy NGC 4889. Its mass is estimated at 8 trillion solar masses and it is thought to be more massive than any other galaxy this close to us. The centre of the galaxy harbours the biggest supermassive black hole that we know of with a mass estimated at around 21 billion solar masses.

It is a mystery how supermassive black holes grew to be so enormous in the limited time since the origin of the universe 13.8 billion years ago, but we know their growth began early. In December 2017 NASA scientists announced the discovery of the most remote quasar ever found. At a distance of over 13 billion light years, the light we are receiving was emitted just 690 million years after the Big Bang. The quasar is generated by an 800-million-solar-mass black hole feasting on the surrounding stars and dust clouds. As yet no one knows how the black hole could have achieved such a size so soon after the birth of the cosmos.

There is also debate about which came first the galaxy or its supermassive black hole. Did galaxies begin with a dramatic round of giant star formation that led rapidly to the creation of supermassive black holes, or did youthful black holes act as seeds for the formation of galaxies?

The next goal of researchers is to image a black hole's event horizon. This is the aim of the Event Horizon Telescope.

Photographing a Black Hole

The ultimate challenge is to actually see a black hole. The easiest target is the supermassive black hole at the centre of our own galaxy, but at 25,000 light years distance even this is beyond the resolving power of the world's best astronomical instruments. Figure 112 shows a computer-generated image of what we might expect to see. The simulation shows a distorted image of the accretion disc whose light is bent around the black hole by the gravitational warping of the surrounding space. The black hole itself appears as a black sphere within the luminous ring.

Shep Doeleman of MIT (Massachusetts Institute of Technology) is leading an ambitious international effort to assemble the Event Horizon Telescope and generate the world's first photograph of a black hole. Success will require at least 5000 times the resolving power of the Hubble Space Telescope. Imaging our galactic super-

Figure 112 A computer-generated image showing what the Event Horizon Telescope is expected to reveal.

massive black hole is comparable to photographing a cricket ball on the Moon. Doeleman has pulled together telescopic resources from around the world in an attempt to achieve this unprecedented resolution.

The galactic centre is shrouded in hot gas which blocks visible light emitted from the stars in this region of the galaxy. Fortunately, infrared radiation is much better at penetrating the murk, so the Event Horizon Telescope is an Earth-sized instrument operating in the far infrared/microwave region of the spectrum. It will combine data collected by a network of radio telescopes around the world to produce images with an unparalleled sharpness. These instruments are located at sites in California, Arizona, Hawaii, Chile, Europe and even the South Pole (Figure 113).

The supermassive black hole at the centre of the giant elliptical galaxy M87 is a second target for the Event Horizon Telescope. It is about 2000 times as distant as the galactic centre, but the diameter of the event horizon of this enormous 6.5-billion-solar-mass black hole is over 1600 times that of the supermassive black hole at the centre of our galaxy. This means that imaging its event

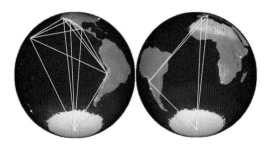

Figure 113 Some of the sites of the telescopes that will form the Event Horizon Telescope. The images from these telescopes will be combined using Very Long Baseline Interferometry.

horizon should be only marginally more difficult. The M87 super-massive black hole is about the size of the orbit of the outermost planet Neptune around the Sun. What's more, it is very active, blasting enormous jets of material into intergalatic space, so it holds out the prospects of revealing even greater secrets.

We are entering a new era for black hole physics. We should have the first direct image of a black hole very soon.

PART III

THE BIOLOGY OF THE COSMOS

The Gorgon's Head!

A great cosmic drama plays out above our heads every night. Perseus, the Greek epic hero, was challenged to visit the land of the hyperboreans, a semi-mythical bleak and frosty land beyond the north wind known today as the island of Britain, his task to retrieve the head of the gorgon Medusa (Figure 114).

The beautiful Medusa had been seduced by the mighty sea god Poseidon and for this indiscretion she was severely punished. The goddess Athena transformed her glorious hair into a nest of serpents and proclaimed a curse that all who gazed on her face would be turned to stone. But Perseus was forewarned and he side-stepped the curse's power. Viewing Medusa's reflected image in his polished bronze shield, with his sword he severed her

Figure 114 *Medusa* by Caravaggio.

The Cosmic Mystery Tour. Nicholas Mee, Oxford University Press (2019). © Nicholas Mee.
DOI: 10.1093/oso/9780198831860.001.0001

snake-wreathed head. At once her bleeding neck disgorged the warrior Chrysaor clad in shining armour and the wingéd horse Pegasus, the offspring of her fatal encounter with the sea god Poseidon. Perseus took up the head of Medusa, mounted Pegasus and rose into the air on course for warmer climes.

On reaching the land of Egypt Perseus encountered a distressing scene. The princess Andromeda was chained to a rock, about to be ravished by a terrible dragon. Andromeda's parents, King Cepheus and Queen Cassiopeia, looked on helplessly, powerless to save the princess. Arriving just in time, Perseus slayed the dragon and saved Andromeda from her dreadful fate (Figure 115).

Six of the most northerly constellations in our night sky are intertwined with this mythic episode. Perseus, Andromeda, the square of Pegasus and the 'W' of Cassiopeia are very prominent features of the night sky. The stars of Cepheus are less obvious and so is the dragon Draco formed of a train of stars close to the north pole. The names for these constellations date back at least two and a half thousand years.

Figure 115 *Perseus Freeing Andromeda* by Piero di Cosimo. A painting with a wonderful air of Monty Python about it.

The Gorgon's Head

Greek philosophers such as Aristotle drew a sharp distinction between the ideal realm of the heavens and the sublunar region of the Earth. Whereas the terrestrial environment was subject to dissolution and decay, the stars were believed to be fixed and unchanging, shining steadily throughout eternity. But these beliefs must have been questioned even in antiquity as not all stars are quite so constant and invariable. In the constellation Perseus lies a curious star that represents the head of Medusa and this identification cannot be coincidental (Figure 116). The star is known as Algol and as with many prominent stars this name has an Arabic origin. Algol derives from Ra's al Ghul, a term used by Islamic astronomers meaning 'the head of the ghoul'. This in turn translates the expression used by Ancient Greek astronomer Ptolemy meaning 'the gorgon's head'. The remarkable feature of

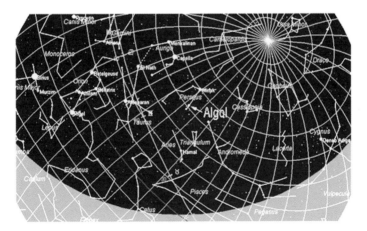

Figure 116 Star map showing the northern constellations Cepheus, Draco, Perseus, Cassiopeia, Andromeda and Pegasus.

Algol that led to this identification is that every three days or so the star fades dramatically for around ten hours and then brightens again. The interval between these dips in brightness is precisely 2 days 20 hours and 49 minutes.

John Goodricke

The strange variability of Algol remained a mystery for many centuries. Not until 1782 did anyone offer an explanation. On 12 November 1782 the 18-year-old John Goodricke (Figure 117) was watching when Algol underwent its curious transformation. As he recorded in his journal, he was astonished by what he saw:

> This night looked at Algol and was much amazed to find its brightness altered. It now appears to be fourth magnitude. I observed it diligently for about an hour upwards hardly believing that it changed its brightness, because I had never heard of any star varying so quick in its brightness. I thought it might be

Figure 117 John Goodricke, painted in 1785 by James Scouler.

perhaps owing to an optical illusion, a defect in my eyes or bad air, but the sequel will show that its change is true and that it was not mistaken.

Goodricke was born in Gröningen in the Netherlands, the son of an English diplomat. At the age of five he contracted scarlet fever and was left profoundly deaf. Fortunately, his wealthy parents were able to send him to specialist schools in Edinburgh and Warrington where he received speech therapy and learnt to lip read. In his later school years he was introduced to astronomy and after leaving school he moved to York where he could observe the night sky with his cousin Edward Pigott in an observatory built by his uncle. It was there that he encountered the monstrous behaviour of Algol. But unlike earlier astronomers who witnessed the uncanny fading of Algol, Goodricke proposed an explanation.

A Spectacular Discovery

William Herschel had made a spectacular discovery just the previous year, 1781, the first new planet in recorded history, known today as Uranus. The existence of a new heavenly body orbiting the Sun in the far reaches of the Solar System may have inspired Goodricke's explanation of Algol's behaviour. Goodricke suggested that Algol might really be a pair of stars orbiting each other so closely they could not be separated even with a telescope. He reasoned that if one of the stars was much brighter than the other, then each time the fainter star passed in front of the brighter star it would block some of its light and this partial eclipse would cause Algol to fade as viewed from Earth, and the regularity of the orbit would produce a clockwork predictability in the dimming events. Goodricke published his theory the following year. It was immediately accepted as the most plausible explanation and he was rewarded with the Copley Medal, the most prestigious award of the Royal Society.

Goodricke's theory is now known to be correct. Algol lies at a distance of 92.8 light years. It consists of a bright star of 3.17 solar masses and a second much fainter star of 0.7 solar masses separated by 9.3 million kilometres, about one-sixteenth of the distance between the Earth and Sun. The two stars orbit each other in just under three days and the regular eclipsing of the brighter star accounts for the periodic dimming of Algol just as Goodricke proposed. Remarkably there is also a third star of 1.76 solar masses in the system, which orbits the inner pair at a distance of around 400 million kilometres and completes one orbit every 680 days, but this does not affect the observed brightness of the stars.

Goodricke continued to study variable stars and found other eclipsing binaries. He also discovered the variability of Delta Cephei, the fourth brightest star in the constellation Cepheus. This was the first discovery of a Cepheid variable, a class of stars that would play an essential role in determining the distances to other galaxies. Goodricke was the rising star of British astronomy. At the age of just 21 on 16 April 1786, he was elected Fellow of the Royal Society. Tragically, he never learnt of this great honour. He died of pneumonia just four days later.

Raise Your Glasses to the Skies!

Science fiction writers have long assumed there are planetary systems around most stars. The stars are so distant, however, that confirming this belief was impossible until relatively recently. But in the past two decades the number of confirmed exoplanets, as they are known, has steadily increased. The most fruitful method for finding them is to search for eclipsing binaries. When a planet crosses the face of a star as viewed from Earth, there is a slight dip in the star's brightness. This is just what happens with Algol and other eclipsing binary stars, but planets are typically far smaller than stars, so they block a much smaller proportion of the star's light. Fortunately, modern detectors, which are essentially sensitive versions of the CCD chips in ordinary digital cameras, can measure very small changes in brightness. If Jupiter passed in front of the Sun, as viewed from outside the Solar System, the gas giant would block about 1% of the light from the Sun. This could easily be detected using a CCD camera. Rocky planets such as Earth and Venus are much smaller and so would be harder to detect. The photograph in Figure 118 shows the 2004 transit of Venus, as viewed from Dorset in England. Venus blocks about 0.01% of the Sun's light.

The Kepler Space Telescope

In 2009 NASA launched the Kepler space observatory (Figure 119), named after the great German astronomer Johannes Kepler, who

The Cosmic Mystery Tour. Nicholas Mee, Oxford University Press (2019). © Nicholas Mee.
DOI: 10.1093/oso/9780198831860.001.0001

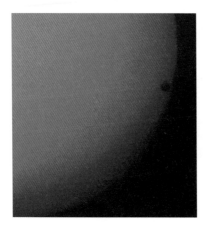

Figure 118 The 2004 Transit of Venus.

first accurately described the orbits of the planets around the Sun. The aim of the mission was to discover planets around distant stars using the eclipsing binary method. Only by chance would a planetary system around another star be oriented in just the right way for the planets to pass in front of the star as viewed from Earth. Nevertheless, by keeping a watchful eye on enough stars we are sure to find some. Kepler continuously monitors the brightness of 145,000 stars and forwards the data to astronomers who perform computerized searches for periodic variations in their brightness.

Kepler has so far discovered over 2500 planets in about 1000 star systems whose existence has been confirmed in follow-up observations, and many other candidates are awaiting confirmation. From these results it is clear that our galaxy contains many billions of rocky planets, some of which must bear a striking resemblance to our own.

A Toast to the Stars!

One of the most interesting planetary systems that we know of is the TRAPPIST-1 system discovered by astronomers at the University

Figure 119 Artist's impression of NASA's space observatory Kepler.

of Liege using the same eclipsing technique. The Belgian project, known as the Transiting Planets and Planetesimals Small Telescope (TRAPPIST), uses a 60-cm robotic telescope at the La Silla Observatory in Chile. Further planets in the same system were discovered in follow-up observations by NASA's Spitzer Space Telescope and the European Southern Observatory's Very Large Telescope at Paranal in Chile. The convoluted acronym TRAPPIST was devised to commemorate Trappist beer, the favourite ale of the Belgian observers, which is still brewed today by the strict Trappist order of Cistercian monks. No doubt the discovery of the TRAPPIST-1 system was celebrated with a glass or two.

TRAPPIST-1 is one of our stellar neighbours at 39 light years distance. It is an ultra-cool red dwarf whose mass is just one-twelfth of the mass of the Sun. The red dwarf is little bigger than Jupiter in size even though it contains almost 100 times Jupiter's mass. Seven planets are known to orbit TRAPPIST-1 and they are labelled 'b' to 'h' in the order they were discovered (Figure 120). The orbits of all the planets would lie well within Mercury's orbit around the Sun, so the TRAPPIST-1 system is like the Solar System in miniature. Planet b orbits the star in just one and a half days, while planet h orbits in 18.8 days. Five of the planets (b, c, e, f, g)

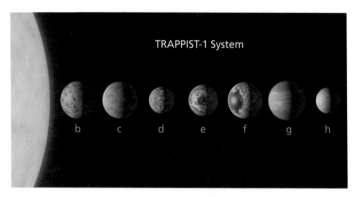

Figure 120 An artist's impression of the seven planets of the TRAPPIST-1 system.

Figure 121 The view from TRAPPIST-1f by Tim Pyle.

are similar in size to the Earth. The other two (d, h) are intermediate in size between Mars and Earth. The view at night from each planet would be spectacular, as they are much closer to each other than the planets in our Solar System, so the other planets in their skies would appear much larger than the Moon does in ours. The red dwarf TRAPPIST-1 is a very small star that emits a tiny fraction of the Sun's radiation. This means that although the planets are so close to the star, the orbits of three of them (e, f, g) are thought to lie in the habitable zone of the system. In this region the temperature would be similar to that on Earth where water is liquid and the existence of life is possible (Figure 121).

Does this mean that we might have neighbours in the TRAPPIST-1 system? Probably not. Although red dwarf stars are much fainter than the Sun, every so often they erupt and emit intense stellar flares from their surface. It is believed that any planet close enough to the red dwarf to engender life would regularly be sterilized by these flares.

Life, But Not as We Know It!

I remember many years ago watching an episode of *Star Trek* called 'The Devil in the Dark'. The Starship *Enterprise* visits a mining community on the planet Janus VI where an unknown lifeform is playing havoc with the mining operations.

The crew eventually track down the source of the disruption—a strange creature that looks like a rather angry pizza (Figure 122).

Figure 122 A silicon-based lifeform surrounded by its rock eggs, as envisaged in *Star Trek*.

Credit: Photo still from *Star Trek*. Courtesy of CBS Television Studios.

The Cosmic Mystery Tour. Nicholas Mee, Oxford University Press (2019). © Nicholas Mee. DOI: 10.1093/oso/9780198831860.001.0001

It is actually a bizarre lifeform unlike anything encountered during their previous adventures. Spock performs a Vulcan mind-meld with the creature and is fascinated to discover that it belongs to a species of sentient beings that live off solid rock and burrow their way through the mountains by secreting very strong acids. The reason for their unusual and rather crunchy diet is that, unlike the familiar carbon-based life on Earth, these are organisms whose fundamental chemistry is based on silicon.

Star Trek's rock munching creature is far from the strangest lifeform dreamt up in science fiction. Here are a couple of my other favourite aliens.

The Black Cloud

The astrophysicist Fred Hoyle wrote a story in 1957 called 'The Black Cloud' in which astronomers discover a vast gas cloud passing through the Solar System. On further investigation this turns out to be a super-intelligent being that is wandering through the galaxy. When scientists work out how to communicate with the cloud, it expresses surprise that there are sentient beings inhabiting a solid planet.

Here Be Dragons

Robert L. Forward went even further in his remarkable novel *Dragon's Egg*, describing the evolution of a civilization on a neutron star. There is no chemistry in the extreme environment of a neutron star, so these organisms are composed of hypothetical complex nuclear structures. This really is life in the ultra-fast lane, as their biology revolves around nuclear physics instead of biochemistry. These fictional nanoscale creatures live their lives a million times faster than us.

Is There Life Out There?

What do we know about the possible existence of life elsewhere? We have a very small sample with which to work, just one planet on which life is known.

Life on Earth is built around a biochemistry based on macro-molecules that have a backbone formed of chains of carbon atoms (Figure 123). Our understanding of chemistry suggests that silicon is not a suitable building block for the complex chemistry required by life. Carbon seems to be uniquely suited to this role, so unfortunately silicon-based lifeforms will probably always remain fiction, for naturally occurring lifeforms anyway. It could be argued that artificial silicon chip-based organisms will some

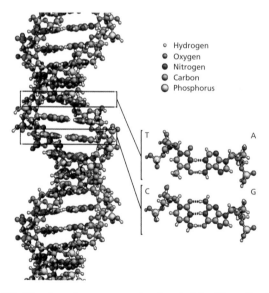

Figure 123 Like the other macromolecules vital for life on Earth, DNA has a backbone of carbon atoms.

day be created. Sentient clouds and nuclear organisms are also probably just entertaining fantasies, but who knows?

Living Is Easy

In all likelihood it is only possible for living organisms to evolve with a carbon-based biochemistry in a liquid water environment. What we do know is that simple lifeforms arose early in the history of our planet. There is evidence for the existence of simple bacterial life as far back as 4 billion years ago, which is around 400 million years after the formation of the Earth, and perhaps as little as 100 million years after the formation of the oceans.

The interval between the arrival of conditions suitable for life and its appearance on Earth seems to have been a remarkably short geological time span. This suggests that life at the bacterial level arises easily and therefore should be common. If this is true, then we could expect to find life elsewhere in the Solar System. There are a number of locations that might be good places to look.

Mars has been on the list for many years. In more recent times attention has focused on satellites of the gas giants, in particular Jupiter's Europa and Saturn's Enceladus.

The God of War

The glaring red eye of Mars has long been associated with war, so it is no surprise that its supposed inhabitants should have been imagined as hostile towards us Earthlings, eager to get their tentacles around our necks (Figure 124).

Although Mars is a cold and hostile world with little atmosphere today, it may have been much more hospitable in the

Figure 124 Illustration by Alvim Corrêa from a 1906 edition of *The War of the Worlds* by H. G. Wells.

distant past. The prospects of finding little green octopuses on Mars disappeared long ago, but there is still some hope of eventually finding microbial lifeforms. There is evidence that water once flowed on Mars, which makes it feasible that life did emerge there. Bacteria survive in even the most challenging environments on Earth, so it is just possible that bacterial life has clung on for billions of years deep within the rock below the surface of Mars.

Europa

Europa is the fourth largest of the Galilean satellites of Jupiter and is a little smaller than our Moon. It is thought to be a good candidate for extraterrestrial life as it appears to have a surface composed of ice that forms a crust over a liquid water ocean. Europa is subjected to a continual flexing and straining by the gravitational pull of the giant planet Jupiter that generates sufficient heat to maintain the liquid ocean. The ice sheets can be seen in Figure 125.

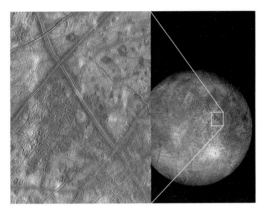

Figure 125 The surface of Europa is covered in ice-sheets. A liquid water ocean is believed to lie beneath the surface.

NASA is currently planning the Europa Clipper Mission. We will learn much more about this icy moon and its possible habitability when it arrives at Europa in the mid 2020s.

Enceladus

Enceladus is a moon of Saturn with a diameter of 500 kilometres. It is covered in fresh uncratered ice. In 2005 the *Cassini* probe discovered plumes of water venting from the south polar region of Enceladus, raising hopes that like Europa it may have a liquid water ocean beneath the surface (Figure 126).

Several other bodies in our Solar System have been suggested as possible homes for living organisms. These include Jupiter's giant moons Ganymede and Callisto, which are also believed to have subsurface oceans; Ceres, the largest of the asteroids; Neptune's large moon Triton; even distant Pluto. Perhaps the most intriguing possibility is Titan.

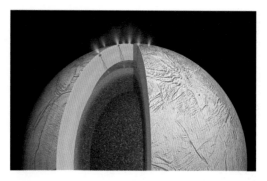

Figure 126 An artist's conception of the plumes of Enceladus.

Titan

The *Cassini* mission to Saturn came to an end when this remarkable space probe took its final plunge into the gas giant on 15 September 2017 after 13 years orbiting the planet. The highlight of *Cassini*'s incredible mission came in 2005 when it released the *Huygens* probe to land on Titan, Saturn's biggest moon (Figure 127). Titan is slightly larger than the planet Mercury. It is one of the few bodies in the Solar System where a probe has touched down. The landing was possible because unlike all the other moons in the Solar System Titan has a thick atmosphere composed mainly of nitrogen and suffused with clouds of methane and ethane.

The temperature on Titan is a rather chilly 180°C below zero. This is far too cold for liquid water to exist on the surface. Nonetheless, Titan has seas, lakes and rivers, but they are composed of methane and ethane, hydrocarbons that play a similar role on Titan to water on Earth. This has led to speculation that Titan might be home to some form of life. It is perhaps rather unlikely, but just about conceivable, that life of some sort could have evolved in this hydrocarbon-rich environment. If so, it must

Figure 127 Artist's conception of the Huygens probe parachuting through the atmosphere of Titan.

be based on an unknown and exotic biochemistry. NASA scientists announced in July 2017 that a chemical known as acrylonitrile has been detected in Titan's atmosphere. This has improved the prospects for finding exotic life there, as it has been suggested that acrylonitrile might be suitable for forming a novel kind of cell membrane.

We should learn much more about Titan in the future. NASA has accepted a proposal for a mission named Dragonfly that will transport a robotic helicopter drone to this intriguing world. Over the course of several years the drone will fly through Titan's atmosphere exploring its varied environments and touching down for extended visits to the most interesting sites. The launch is currently scheduled for 2024 or 2025.

Finding any sort of life elsewhere in the Solar System would be sensational. It would suggest that Earth is not just an incredibly lucky one-off and that life is abundant throughout the universe.

To Boldly Go …

We Earthlings have long dreamed of visiting the stars. This fantasy may soon become a reality—remotely at least.

The first artificial craft ever sent beyond the Solar System was *Pioneer 10*, launched in March 1972 (Figure 128). Its main mission was to study the planet Jupiter and it was the first probe to visit such a distant world. *Pioneer 10*'s stunning close-up images of the gas giant helped to transform our understanding of the Solar System.

Figure 128 *Pioneer 10* during final preparations.

The Cosmic Mystery Tour. Nicholas Mee, Oxford University Press (2019). © Nicholas Mee.
DOI: 10.1093/oso/9780198831860.001.0001

Swinging past Jupiter in December 1973, *Pioneer 10* was propelled beyond the gravitational clutches of the Sun and out of the Solar System. The craft continues on its way in the direction of the star Aldebaran, the bright red eye of the angry bull that is the constellation Taurus. Travelling at around 40,000 kilometres per hour, it will take *Pioneer 10* about two million years to cover the 65 light years separating us from Aldebaran.

Carl Sagan and Frank Drake, pioneers in the search for extraterrestrial intelligence, created a striking design that is engraved on a gold-plated plaque carried by *Pioneer 10* (Figure 129). The plaque will offer clues to the origin of the probe in the extremely unlikely event that it is ever found by an alien intelligence. The design shows the trajectory of the probe through the Solar System and gives a key to indicate where the probe came from by mapping the directions from the Sun to various pulsars. It was assumed that any civilization sophisticated enough to intercept an interstellar visitor would easily decipher the symbols.

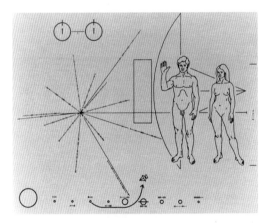

Figure 129 The *Pioneer 10* plaque designed by Carl Sagan and Frank Drake. (Judging by the probe's size in Figure 128, NASA seems to have wanted the aliens to believe it was built by giants.)

The plaque shows the shape and size of the creatures that dispatched the spacecraft by comparison with the size of the probe. This also gives us an idea of just how hefty a craft *Pioneer 10* was. In the decades since *Pioneer* there have been incredible developments in technology. In particular, computer processors and other solid state devices have shrunk until chips of a few grammes in weight that sit comfortably in the palm of one hand contain far more processing power than any probe of the 1970s.

Breakthrough Starshot

In April 2016 an ambitious research project Breakthrough Starshot was launched with the aim of developing technologies that will enable us to visit the stars. The serious nature of the project is clear from the profile of its board members. These included world-renowned physicist Stephen Hawking, famous for his revolutionary ideas about black holes, who has now sadly died, and Freeman Dyson, one of the architects of the quantum theory of electromagnetism (QED) (Figure 130). Dyson was also a leading figure in

Figure 130 Stephen Hawking and Freeman Dyson launching Breakthrough Starshot.

Project Orion, a quite staggering secret project sponsored by the United States government in the early 1960s. Its goal was to build spaceships that would ride shock waves generated by trails of small nuclear bombs dispensed and detonated from the rear of the spacecraft. Dyson was planning to send astronauts to Saturn well before Apollo reached the Moon. The board also includes tech billionaire Mark Zuckerberg, founder of Facebook, and Russian billionaire Yuri Milner, who has donated $100 million of development funding to kickstart the project.

Breakthrough Starshot aims to show the feasibility of sending a mission to our nearest stellar neighbour, Alpha Centauri, which lies at a distance of 4.37 light years or 40 trillion kilometres. Traversing such an immense void is anything but straightforward and there are many technological challenges that must be overcome if the project is to succeed, but the team is optimistic that the mission can be launched within about twenty years.

Of course, there won't be any people on board. The aim is to send a miniature nanocraft, essentially a silicon chip containing a computer, camera, laser communication system and a plutonium power source, attached to a four-metre by four-metre sail (Figure 131). The propulsion will come from an array of Earth-based

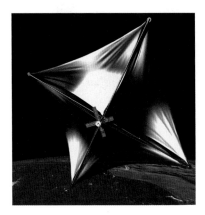

Figure 131 A spaceship as envisaged by Breakthrough Starshot.

Figure 132 An artist's impression of the laser array for Breakthrough Starshot situated at high altitude in a location such as the Atacama Desert in Chile.

lasers firing a 100-gigawatt beam at the sails (Figure 132). Within a matter of minutes this will accelerate the craft to one-fifth of the speed of light. It will cross Mars' orbit within an hour and, whereas the recent *New Horizons* probe took nine years to reach Pluto, the nanoprobe will cross the orbit of Pluto the day after launch. Like the schooners of old sailing the world's oceans, these miniature ships will be blown to the stars across oceans of interstellar space.

Alpha Centauri is a multiple-star system. It has two main components that orbit each other every 80 years. They are designated Alpha Centauri A and B and have 1.1 and 0.9 solar masses, respectively. At their closest they are separated by a distance comparable to that between the Sun and Saturn. A third star known as Proxima Centauri, because at 4.24 light years it is the closest star to us, orbits the two main stars every half a million years or so. Proxima is a very faint red dwarf with just one-eighth the mass of the Sun. In August 2016 the European Southern Observatory announced the discovery of a planet orbiting Proxima with a slightly greater mass than the Earth. This planet, which lies in the habitable zone and completes one orbit every eleven days, will definitely put Proxima on the tourist route to the main stars of Alpha Centauri.

At one-fifth the speed of light the journey to Alpha Centauri will take just over twenty years. The nanocraft will then send back images and information about the stars and planets in the system and it will take over four years for us to receive this data. There is no means of slowing down the nanocraft on arrival, so it will race past at ultrahigh speed. To overcome this constraint the plan is to release a fleet of hundreds of such craft at short intervals. A mothership containing the nanocraft will be launched into Earth orbit and then release the nanocraft one by one. Each nanocraft will use photon thrusters to position itself in the path of the ground-based laser system, then hoist the main brace and await a fair wind before being blasted to the stars. Twenty years or so later the nanocraft will race through the Alpha Centauri system in a single day, but if 100 nanocraft are launched at daily intervals, as one exits the star system the next will enter and a continuous stream of data will be beamed back to Earth for 100 days.

Many daunting hurdles must be overcome before sailing to the stars becomes a reality, but the technological advances that will be made along the way should benefit us all.

Somewhere over the Rainbow

In 1960, the astronomer Frank Drake weighed up the options for his first attempt to find extraterrestrial intelligence. Two nearby stars Epsilon Eridani and Tau Ceti, 10 and 12 light years distant, respectively, seemed like promising candidates. Both are similar to the Sun with around four-fifths of the Sun's mass. They were ideal targets for Drake's quest for alien civilizations, Project Ozma, named after the queen of the land over the rainbow in the books of L. Frank Baum. Nothing was known about any planetary systems that Epsilon Eridani and Tau Ceti might have; thirty years would pass before the first discovery of an exoplanet in any system. Tau Ceti is now thought to have a retinue of at least five planets with masses ranging from two to six Earths. Drake pointed the 26-metre radio telescope at Green Bank, West Virginia towards each of the two stars with great anticipation and listened for anything unusual that might indicate the presence of alien communications (Figure 133). Nothing was heard. This was the beginning of SETI (the search for extraterrestrial intelligence).

Follow the Yellow Brick Road

Undeterred, Drake attempted to estimate the number of alien civilizations in our galaxy to rally support for a wider search. As Drake was well aware, determining the likely evolution of intelligent beings on other worlds is an almost hopeless task given our limited knowledge of fundamental questions such as the origin of life and the nature of intelligence. Nonetheless, he set these difficulties aside

The Cosmic Mystery Tour. Nicholas Mee, Oxford University Press (2019). © Nicholas Mee.
DOI: 10.1093/oso/9780198831860.001.0001

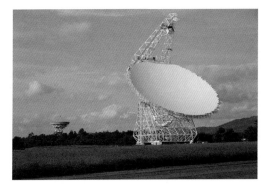

Figure 133 The Green Bank Radio Observatory.

and devised a formula, now known as the *Drake equation*, for calculating the number of detectable civilizations. It combines all the conditions that are necessary for alien civilizations to develop. Even today, sixty years later, we remain in the dark about several of the terms. But Drake believed the equation would focus attention on where our ignorance lies, which would ultimately help to resolve these issues.

The numbers that Drake required to determine whether SETI might succeed are the following (Figure 134):

R_*—the average rate of star formation in our galaxy

f_p—the fraction of stars that have planets

n_e—the proportion of planets that could potentially support life

f_l—the fraction of those planets on which life actually arises

f_i—the fraction of life-bearing planets where intelligent civilizations evolve

f_c—the fraction of these civilizations that develop communications technologies, such as radio, that are detectable from other star systems

L—the length of time for which such civilizations survive and emit detectable signals

Figure 134 A colourful representation of the Drake equation.

When multiplied together these terms give a figure N representing the current number of alien civilizations in our galaxy whose signals we might hope to detect:

$$N = R_* \cdot f_p \cdot n_e \cdot f_l \cdot f_i \cdot f_c \cdot L.$$

Drake knew there was great uncertainty in most terms in his equation, but he was keen to show there was good reason to continue the search for extraterrestrial signals. He convened a meeting of distinguished scientists at Green Bank Observatory with expertise in all the diverse fields that were relevant to the equation. After considering each term, the team concluded that given current knowledge of astrophysics and the evolution of life, their best estimate was that the galaxy should be home to around 10,000 advanced civilizations; therefore, a serious search for extraterrestrial intelligence stood a real chance of success. Drake reckoned that if around ten million stars were monitored, eventually detection was assured, although this might take decades or even centuries. At the end of the conference the delegates raised their champagne glasses and Otto Struve, the director of Green Bank, offered up a toast:

'To the value of L. May it prove to be a very large number.'

Finding Your Very Own Alien

In the decades since the Green Bank conference there have been ever more sophisticated attempts to scan the heavens with radio telescopes on the lookout for signals from extraterrestrials. Much of this work is now coordinated by the SETI Institute based in California. Project Phoenix, which began in 1995, has used some of the world's largest radio telescopes at Green Bank, Arecibo in Puerto Rico and the Parkes Observatory in Australia to target 800 Sun-like stars within a range of 200 light years for signs of alien communications. So much data has been collected that processing the data is now the bottleneck. To tackle this issue the SETI@Home project was set up in 1999. It has become one of the biggest distributed computing projects in the world. Everyone is welcome to get involved by visiting the website: https://setiathome.berkeley.edu/. Once the SETI@Home software is installed, any computer downtime will be utilized to analyse data from Arecibo and Green Bank while a screensaver runs in the foreground (Figure 135).

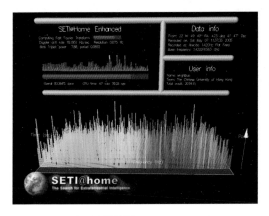

Figure 135 The SETI@Home screensaver.

Is There Anybody Out There?

The time available for SETI research is quite limited at the long-established observatories. But now a dedicated radio observatory, the Allen Telescope Array (ATA), is under construction. The ATA is named after Microsoft co-founder Paul Allen, who is funding the project. It is intended to perform full sky surveys and will combine pure astronomy research with SETI. The array began operations in 2007 with 42 six-metre radio dishes and when complete there will be 350 dishes. The ATA will increase the targeted analysis of star systems by a factor of at least 100. It is hoped that over the next two decades stellar reconnaissance will increase to a million or more nearby stars deemed possible hosts of life-bearing planets.

In the years since Drake devised his equation there is at least one relevant area where our knowledge has advanced significantly. We now know that most stars have planetary systems and the results from the Kepler Space Telescope suggest the galaxy contains at least 40 billion rocky planets. So there is no shortage of planetary real estate.

With so many rocky planets available we might assume that life is abundant in the galaxy and that technological civilizations abound. But is this really true? The great nuclear physicist Enrico Fermi thought not. He argued that it could not be so, but that is a story for the next chapter.

Where Is Everybody?

Enrico Fermi was born in Rome in 1901 (Figure 136). He was one of the great physicists of the twentieth century and made many important contributions to physics including a ground-breaking theory of the weak nuclear force that predicted the existence of neutrinos. He was also a leading experimental physicist and created the first nuclear reactor under the sports fields at the University of Chicago in December 1942. The artificial element number 100 fermium is named in his honour.

Figure 136 Enrico Fermi.

The Cosmic Mystery Tour. Nicholas Mee, Oxford University Press (2019). © Nicholas Mee.
DOI: 10.1093/oso/9780198831860.001.0001

One day while working at the nuclear research centre at Los Alamos in 1950 Fermi strolled to lunch with his colleagues Emil Konopinski, Edward Teller and Herbert York. Press speculation about alien spacecraft was rife at the time and the conversation turned to recent reports of UFO sightings and a newspaper cartoon blaming aliens for the theft of municipal rubbish bins (Figure 137). Later, during lunch Fermi suddenly exclaimed: 'Where Is Everybody?' Teller recalled long afterwards that this was met with roars of laughter. Everyone knew what Fermi meant, but his startling question seemed to come from nowhere. Fermi followed up his lunchtime observation with a series of calculations estimating the probability that technologically advanced civilizations would arise and concluded that we ought to have been visited long ago and many times over. The lack of evidence for such encounters has become known as the *Fermi paradox*.

Figure 137 The aliens and the municipal trashcans.

The Fermi Paradox

Crude inferences based on the vast number of stars in our galaxy (Figure 138) might suggest that plenty of intelligent lifeforms must be out there somewhere. But according to Fermi this might not be the case. His argument boils down to the following observations. The first stars in the galaxy appeared soon after the Big Bang 13.8 billion years ago. There would have been a long time lag in setting the stage for the arrival of sentient beings. Billions of years would pass while carbon, nitrogen, oxygen and the other elements of life accumulated in quantities sufficient for life-bearing planets to form. But the Earth is a mere 4.57 billion years old and it has taken all of this 4.57 billion years for a technological civilization to evolve. Many stars are older than the Sun and the evolution of intelligence on Earth was a long and slow process. There is every reason to suppose therefore that

Figure 138 The Milky Way as seen from northern Chile.

intelligent life could have appeared on other worlds long before ours. If the conditions for the evolution of intelligent life are prevalent throughout the galaxy, then we should expect that technological civilizations have arisen millions if not billions of years ago.

It is just 5500 years since the earliest writing in Ancient Sumeria and Egypt, 400 years since the first telescope, 120 years since the first radio communications and 30 years since the dawn of the World Wide Web. The whole of human technology has arisen within a brief few thousand years, so we must expect the technology of any ancient alien civilization to have long surpassed our own. Nevertheless, as we saw in an earlier chapter, we are already contemplating the possibility of sending nanoprobes to the stars, as proposed in Breakthrough Starshot. It is only natural that other civilizations are equally curious and have similar desires to explore the stars. But we can barely imagine the capabilities of a civilization 1000 years in advance of our own, never mind one a million years in advance. It seems reasonable to suppose that such a civilization would have developed fleets of robotic probes that could colonize planets in distant star systems and utilize the resources of those planets to build further fleets of robotic probes that could seek out new star systems to explore and colonize.

The diameter of our galaxy is around 100,000 light years. At one-fifth of the speed of light, the speed envisaged by Breakthrough Starshot, it would take no more than half a million years to cross the entire galaxy. There is no reason why, in principle, fleets of self-replicating robotic probes could not colonize the entire galaxy within one or at most a few million years, and, as Fermi pointed out, if the galaxy really does harbour technological civilizations, then they should have been around for much longer than this. Therefore, the colonization of the galaxy should already have happened. So, as Fermi said: 'Where Is Everybody?'

Self-destruction

A number of answers have been put forward to explain the Fermi paradox. One disheartening possibility is that technological civilizations do frequently arise, but they are short-lived as their technology inevitably leads to self-destruction and the rapid demise of the civilization. This was a concern of the delegates who met to discuss the search for extraterrestrial intelligence (SETI) at Green Bank observatory in 1961. This was at the height of the Cold War, less than a year before the Cuban Missile Crisis, which threatened humanity with a return to the Stone Age and possibly even total annihilation (Figure 139). The threat of nuclear holocaust may have receded somewhat, but we now have global warming and ecological degradation to contend with. Who knows whether we will survive long enough to colonize the galaxy?

Figure 139 The first hydrogen bomb test, code-named Ivy Mike.

The Zoo Hypothesis

For the more optimistic there are other explanations such as the possibility that they are here, but we haven't noticed! Perhaps we have been visited by stealth nanobots, but the extraterrestrials choose not to disclose their presence. Explanations of this sort go by the name of the zoo hypothesis.

We Really Are the First!

Then there is the daunting possibility that there are no other alien civilizations and we really are the first technological civilization to evolve in the Milky Way Galaxy. This startling proposition is sometimes known as the rare Earth hypothesis.

In the words of Sir Arthur C. Clarke:

Two possibilities exist: either we are alone in the universe or we are not. Both are equally terrifying.

Further Reading

There is a lot more about the structure of matter and particle physics in my book: Nicholas Mee, *Higgs Force: Cosmic Symmetry Shattered* (Quantum Wave, 2012).

For more about gravity, from the earliest cosmologies to Einstein's theory of general relativity and supermassive black holes, see my book: Nicholas Mee, *Gravity: Cracking the Cosmic Code* (Virtual Image, 2014).

I regularly add posts about a wide range of science and maths topics to my blog at: http://quantumwavepublishing.co.uk/blog/

The Path to Immortality

Richard Westfall, *Never at Rest: A Biography of Isaac Newton* (Cambridge University Press, 1980). This is the definitive account of Newton's life and incredible achievements. Newton's activities at the Royal Mint are to be found in Chapter 12: The Mint.

The Rosetta Stone and Quantum Waves

Julian Voss-Andreae's remarkable sculptures can be seen at: http://www.JulianVossAndreae.com/

Jim Al-Khalili, *Quantum: A Guide for the Perplexed* (Weidenfeld & Nicholson, 2012). This is a very readable account of the strange world of quantum mechanics.

There is more about quantum mechanics and its technological applications in: Tony Hey and Patrick Walters, *The New Quantum Universe* (Cambridge University Press, 2003).

Cosmic Ripples

The latest news about LIGO and the detection of gravitational waves is available at: https://www.ligo.caltech.edu/

Lovely LISA

For more about the proposed space-based gravitational wave detector LISA, visit the website of the LISA consortium at: https://www.elisascience.org/. Further details are available on the ESA website: http://sci.esa.int/lisa/

Animated Atom Boy

IBM's incredible atomic scale video *A Boy and His Atom: The World's Smallest Movie* can be viewed here: https://www.youtube.com/watch?v=oSCX78-8-q0

Forces of the World Unite!

The latest news about research at the Large Hadron Collider is available at: https://home.cern/about/updates

Most of the Universe is Missing!

Further details about the XENON Dark Matter Project in the Gran Sasso laboratory can be found at: http://xenon.astro.columbia.edu/XENON100_Experiment/

To keep up-to-date with developments in the search for dark matter around the world visit: http://www.interactions.org/hub/dark-matter-hub

The Battle for the Cosmos!

The discovery and significance of the cosmic microwave background are the subject of: George Smoot, *Wrinkles in Time: Witness to the Birth of the Universe* (Harper Perennial Reprint Edition, 2009).

Alchemical Furnaces of the Cosmos

An engaging description of the synthesis of the elements in the stars can be found in: Ken Croswell, *The Alchemy of the Heavens: Searching for Meaning in the Milky Way* (Anchor, 1996).

Diamonds in the Sky!

Tracking down the white dwarf Sirius B is notoriously difficult. If you have access to a good telescope and you would like to give it a go, the following web page might help: http://www.skyandtelescope.com/observing/sirius-b-a-new-pup-in-my-life/

From the Leviathan to the Behemoth

More information about the European Southern Observatory and its telescopes is available on ESO's website: http://www.eso.org. ESO is currently building The Extremely Large Telescope. You can keep up-to-date with its progress here: http://www.eelt.org.uk/

The Ultimate Heavy Metal Space Rock

If you are curious about what a pulsar might sound like, take a listen to the recordings on the Jodrell Bank website: http://www.jb.man.ac.uk/pulsar/Education/Sounds/

Pan Galactic Gargle Blaster

There is more about the Hulse-Taylor binary neutron star system at: http://www.astro.cornell.edu/academics/courses/astro201/psr1913.htm

Doctor Atomic and the Black Hole

Richard Rhodes, *The Making of the Atomic Bomb* (Simon and Schuster, 1986). This is one of the best popular science books ever written and includes the story of Robert Oppenheimer and the Manhattan Project.

Kip Thorne, *Black Holes and Time Warps: Einstein's Outrageous Legacy* (Papermac, 1995). This is an engaging description of general relativity and its implications from one of the world's leading gravity researchers.

Supermassive Black Holes

Mitchel Begelman and Martin Rees, *Gravity's Fatal Attraction: Black Holes in the Universe* (Cambridge University Press, 2009). This is a very readable book about black holes and how we came to understand them.

For more about the ongoing investigations into the supermassive black hole at the centre of our galaxy, visit UCLA's Galactic Center Group at: http://www.galacticcenter.astro.ucla.edu/science.html

The official website of the Event Horizon Telescope is at: http://www.eventhorizontelescope.org

Raise Your Glasses to the Skies!

The current tally of planets discovered by the Kepler Space Telescope can be found at: https://www.nasa.gov/mission_pages/kepler/main/index.html

There is much more information about the TRAPPIST-1 planetary system including some very attractive illustrations at: http://www.trappist.one/

Life, But Not as We Know It!

Fred Hoyle's story *The Black Cloud* is discussed on the following Wikipedia page: https://en.wikipedia.org/wiki/The_Black_Cloud

Robert L. Forward, *Dragon's Egg* (Ballantine, 1980). This is one of the most imaginative science fiction stories of alien life ever written.

There is more about NASA's planned Europa Clipper mission at: https://www.jpl.nasa.gov/missions/europa-clipper. For more about the proposed Dragonfly mission to Titan see: http://dragonfly.jhuapl.edu/

To Boldly Go . . .

The official website of Breakthrough Starshot, the incredibly ambitious project to send a probe to the nearest stars, can be found here: https://breakthroughinitiatives.org/Initiative/3

Somewhere over the Rainbow

The SETI Institute is available at: https://www.seti.org, with details about the Allen Telescope Array here: https://www.seti.org/ata. If you would like to sign up to SETI@Home visit: https://setiathome.berkeley.edu/

Where Is Everybody?

Peter Ward and Donald Brownlee, *Rare Earth: Why Complex Life is Uncommon in the Universe* (Springer, 2009). This is a detailed discussion of what makes the Earth exceptional and why intelligent civilizations might be extremely rare.

Illustration Credits

Part I opener: Nicholas Mee; Fig. 1: Wikimedia - Bob Collowan/Commons/ CC-BY-SA-4.0; Fig. 2: Wikimedia—Cmglee; Fig. 4: ESA; Fig. 5: ESA/NASA/ HST; Fig. 6: The Royal Mint; Fig. 8: Wikimedia—Hedwig Storch; Fig. 9: PASCO Scientific; Fig. 10: Julian Voss-Andreae, *Quantum Man*, 2006, Steel with patina, 100" × 44" × 20" (2.50 × 1.10 × 0.50 m), Location: City of Moses Lake, Washington, USA; Fig. 11: Dartmouth College Electron Microscope Facility; Fig. 12: CERN, Geneva; Fig. 14: Nicholas Mee; Fig. 16: Nicholas Mee; Fig. 17: Nicholas Mee; Fig. 18: John Jenkins; Fig. 19: Nicholas Mee; Fig. 22: LIGO Scientific Collaboration (LSC) / NASA; Fig. 23: Caltech/MIT/ LIGO Lab; Fig. 24: LIGO/T.Pyle; Fig. 25: Caltech/MIT/LIGO Lab; Fig. 26: LIGO/Caltech/MIT/Sonoma State (Aurore Simonnet); Fig. 27: NASA; Fig. 28: ESA/ATG medialab; Fig. 29: ESO/L. Calçada; Fig. 30: Hellenic Post; Fig. 31: M. F. Crommie, C. P. Lutz and D. M. Eigler, IBM Research Division, Almaden Research Center, California; Fig. 32: Don Eigler, IBM Research Division, Almaden Research Center, California; Fig. 33: *A Boy and His Atom*, IBM—Andreas Heinrich; Fig. 34: Nicholas Mee; Fig. 35: Oak Ridge Associated Universities; Fig. 36: NASA; Fig. 37: Nicholas Mee; Fig. 38: PostNord; Fig. 39: Nicholas Mee; Fig. 40: Todd Helmenstine—*Science Notes* http://sciencenotes.org/printable-periodic-table/; Fig. 42: Wikimedia - Geni; Fig. 44: Nicholas Mee; Fig. 45: Victor Stabin, Feynman stamp, US Postal Service; Fig. 46: CERN, Geneva; Fig. 48: Nicholas Mee; Fig. 49: Nicholas Mee; Fig. 50: NASA; Fig. 51: ESO Fig. 52: XENON - Wikimedia - Jpiemaar13; Fig. 53: X-ray: NASA/CXC/CfA/ M.Markevitch et al. / Lensing Map: NASA/ STScI; ESO WFI; Magellan/U.Arizona/ D.Clowe et al. / Optical: NASA/STScI; Magellan/U.Arizona/D.Clowe et al.; Part II opener: Illustris Collaboration/ Illustris Simulation; Fig. 55: Opera North; Fig. 56: BPost; Fig. 57: *Womb* by John Robinson and Nicholas Mee; Fig. 58: Christopher S. Baird; Fig. 59: Wikimedia; Fig. 60: Bob King—Astro Bob; Fig. 61: George Gamow, edited by Russell Stannard, illustrated by Michael Edwards, The New World of Mr Tompkins, (1999) © Cambridge University Press 1999, reproduced with permission; Fig. 62: *Newsweek*; Fig. 63: *Genesis of the Big Bang* (Oxford University Press, 2001), by Ralph Alpher and Robert Herman; Fig. 64: ESA/Hubble;

Index

Page numbers in *italics* refer to illustrations.